RILEM State-of-the-Art Reports

RILEM STATE-OF-THE-ART REPORTS
Volume 33

RILEM, The International Union of Laboratories and Experts in Construction Materials, Systems and Structures, founded in 1947, is a non-governmental scientific association whose goal is to contribute to progress in the construction sciences, techniques and industries, essentially by means of the communication it fosters between research and practice. RILEM's focus is on construction materials and their use in building and civil engineering structures, covering all phases of the building process from manufacture to use and recycling of materials. More information on RILEM and its previous publications can be found on www.RILEM.net. The RILEM State-of-the-Art Reports (STAR) are produced by the Technical Committees. They represent one of the most important outputs that RILEM generates – high level scientific and engineering reports that provide cutting edge knowledge in a given field. The work of the TCs is one of RILEM's key functions. Members of a TC are experts in their field and give their time freely to share their expertise. As a result, the broader scientific community benefits greatly from RILEM's activities. RILEM's stated objective is to disseminate this information as widely as possible to the scientific community. RILEM therefore considers the STAR reports of its TCs as of highest importance, and encourages their publication whenever possible. The information in this and similar reports is mostly pre-normative in the sense that it provides the underlying scientific fundamentals on which standards and codes of practice are based. Without such a solid scientific basis, construction practice will be less than efficient or economical. It is RILEM's hope that this information will be of wide use to the scientific community.

Indexed in SCOPUS, Google Scholar and SpringerLink.

More information about this series at http://www.springer.com/series/8780

Jorge Branco · Philipp Dietsch ·
Thomas Tannert

Editors

Reinforcement of Timber Elements in Existing Structures

State-of-the-Art Report of the RILEM TC
245-RTE

 Springer

Editors
Jorge Branco
Department of Civil Engineering
University of Minho
Guimarães, Portugal

Thomas Tannert
University of Northern British Columbia
Prince George, British Columbia, Canada

Philipp Dietsch
Unit of Timber Engineering
Department of Structural Engineering
and Material Sciences
University of Innsbruck
Innsbruck, Austria

ISSN 2213-204X   ISSN 2213-2031   (electronic)
RILEM State-of-the-Art Reports
ISBN 978-3-030-67796-1   ISBN 978-3-030-67794-7   (eBook)
https://doi.org/10.1007/978-3-030-67794-7

This Springer imprint is published by the registered company Springer Nature Switzerland AG
The registered company address is: Gewerbestrasse 11, 6330 Cham, Switzerland

RILEM Publications

The following list is presenting the global offer of RILEM Publications, sorted by series. Each publication is available in printed version and/or in online version.

RILEM Proceedings (PRO)

PRO 1: Durability of High Performance Concrete (ISBN: 2-912143-03-9; e-ISBN: 2-351580-12-5; e-ISBN: 2351580125); *Ed. H. Sommer*

PRO 2: Chloride Penetration into Concrete (ISBN: 2-912143-00-04; e-ISBN: 2912143454); *Eds. L.-O. Nilsson and J.-P. Ollivier*

PRO 3: Evaluation and Strengthening of Existing Masonry Structures (ISBN: 2-912143-02-0; e-ISBN: 2351580141); *Eds. L. Binda and C. Modena*

PRO 4: Concrete: From Material to Structure (ISBN: 2-912143-04-7; e-ISBN: 2351580206); *Eds. J.-P. Bournazel and Y. Malier*

PRO 5: The Role of Admixtures in High Performance Concrete (ISBN: 2-912143-05-5; e-ISBN: 2351580214); *Eds. J. G. Cabrera and R. Rivera-Villarreal*

PRO 6: High Performance Fiber Reinforced Cement Composites—HPFRCC 3 (ISBN: 2-912143-06-3; e-ISBN: 2351580222); *Eds. H. W. Reinhardt and A. E. Naaman*

PRO 7: 1st International RILEM Symposium on Self-Compacting Concrete (ISBN: 2-912143-09-8; e-ISBN: 2912143721); *Eds. Å. Skarendahl and Ö. Petersson*

PRO 8: International RILEM Symposium on Timber Engineering (ISBN: 2-912143-10-1; e-ISBN: 2351580230); *Ed. L. Boström*

PRO 9: 2nd International RILEM Symposium on Adhesion between Polymers and Concrete ISAP '99 (ISBN: 2-912143-11-X; e-ISBN: 2351580249); *Eds. Y. Ohama and M. Puterman*

PRO 10: 3rd International RILEM Symposium on Durability of Building and Construction Sealants (ISBN: 2-912143-13-6; e-ISBN: 2351580257); *Ed. A. T. Wolf*

PRO 11: 4th International RILEM Conference on Reflective Cracking in Pavements (ISBN: 2-912143-14-4; e-ISBN: 2351580265); *Eds. A. O. Abd El Halim, D. A. Taylor and El H. H. Mohamed*

PRO 12: International RILEM Workshop on Historic Mortars: Characteristics and Tests (ISBN: 2-912143-15-2; e-ISBN: 2351580273); *Eds. P. Bartos, C. Groot and J. J. Hughes*

PRO 13: 2nd International RILEM Symposium on Hydration and Setting (ISBN: 2-912143-16-0; e-ISBN: 2351580281); *Ed. A. Nonat*

PRO 14: Integrated Life-Cycle Design of Materials and Structures—ILCDES 2000 (ISBN: 951-758-408-3; e-ISBN: 235158029X); (ISSN: 0356-9403); *Ed. S. Sarja*

PRO 15: Fifth RILEM Symposium on Fibre-Reinforced Concretes (FRC)—BEFIB'2000 (ISBN: 2-912143-18-7; e-ISBN: 291214373X); *Eds. P. Rossi and G. Chanvillard*

PRO 16: Life Prediction and Management of Concrete Structures (ISBN: 2-912143-19-5; e-ISBN: 2351580303); *Ed. D. Naus*

PRO 17: Shrinkage of Concrete—Shrinkage 2000 (ISBN: 2-912143-20-9; e-ISBN: 2351580311); *Eds. V. Baroghel-Bouny and P.-C. Aïtcin*

PRO 18: Measurement and Interpretation of the On-Site Corrosion Rate (ISBN: 2-912143-21-7; e-ISBN: 235158032X); *Eds. C. Andrade, C. Alonso, J. Fullea, J. Polimon and J. Rodriguez*

PRO 19: Testing and Modelling the Chloride Ingress into Concrete (ISBN: 2-912143-22-5; e-ISBN: 2351580338); *Eds. C. Andrade and J. Kropp*

PRO 20: 1st International RILEM Workshop on Microbial Impacts on Building Materials (CD 02) (e-ISBN 978-2-35158-013-4); *Ed. M. Ribas Silva*

PRO 21: International RILEM Symposium on Connections between Steel and Concrete (ISBN: 2-912143-25-X; e-ISBN: 2351580346); *Ed. R. Eligehausen*

PRO 22: International RILEM Symposium on Joints in Timber Structures (ISBN: 2-912143-28-4; e-ISBN: 2351580354); *Eds. S. Aicher and H.-W. Reinhardt*

PRO 23: International RILEM Conference on Early Age Cracking in Cementitious Systems (ISBN: 2-912143-29-2; e-ISBN: 2351580362); *Eds. K. Kovler and A. Bentur*

PRO 24: 2nd International RILEM Workshop on Frost Resistance of Concrete (ISBN: 2-912143-30-6; e-ISBN: 2351580370); *Eds. M. J. Setzer, R. Auberg and H.-J. Keck*

PRO 25: International RILEM Workshop on Frost Damage in Concrete (ISBN: 2-912143-31-4; e-ISBN: 2351580389); *Eds. D. J. Janssen, M. J. Setzer and M. B. Snyder*

PRO 26: International RILEM Workshop on On-Site Control and Evaluation of Masonry Structures (ISBN: 2-912143-34-9; e-ISBN: 2351580141); *Eds. L. Binda and R. C. de Vekey*

PRO 27: International RILEM Symposium on Building Joint Sealants (CD03; e-ISBN: 235158015X); *Ed. A. T. Wolf*

PRO 28: 6th International RILEM Symposium on Performance Testing and Evaluation of Bituminous Materials—PTEBM'03 (ISBN: 2-912143-35-7; e-ISBN: 978-2-912143-77-8); *Ed. M. N. Partl*

PRO 29: 2nd International RILEM Workshop on Life Prediction and Ageing Management of Concrete Structures (ISBN: 2-912143-36-5; e-ISBN: 2912143780); *Ed. D. J. Naus*

PRO 30: 4th International RILEM Workshop on High Performance Fiber Reinforced Cement Composites—HPFRCC 4 (ISBN: 2-912143-37-3; e-ISBN: 2912143799); *Eds. A. E. Naaman and H. W. Reinhardt*

PRO 31: International RILEM Workshop on Test and Design Methods for Steel Fibre Reinforced Concrete: Background and Experiences (ISBN: 2-912143-38-1; e-ISBN: 2351580168); *Eds. B. Schnütgen and L. Vandewalle*

PRO 32: International Conference on Advances in Concrete and Structures 2 vol. (ISBN (set): 2-912143-41-1; e-ISBN: 2351580176); *Eds. Ying-shu Yuan, Surendra P. Shah and Heng-lin Lü*

PRO 33: 3rd International Symposium on Self-Compacting Concrete (ISBN: 2-912143-42-X; e-ISBN: 2912143713); *Eds. Ó. Wallevik and I. Nielsson*

PRO 34: International RILEM Conference on Microbial Impact on Building Materials (ISBN: 2-912143-43-8; e-ISBN: 2351580184); *Ed. M. Ribas Silva*

PRO 35: International RILEM TC 186-ISA on Internal Sulfate Attack and Delayed Ettringite Formation (ISBN: 2-912143-44-6; e-ISBN: 2912143802); *Eds. K. Scrivener and J. Skalny*

PRO 36: International RILEM Symposium on Concrete Science and Engineering —A Tribute to Arnon Bentur (ISBN: 2-912143-46-2; e-ISBN: 2912143586); *Eds. K. Kovler, J. Marchand, S. Mindess and J. Weiss*

PRO 37: 5th International RILEM Conference on Cracking in Pavements—Mitigation, Risk Assessment and Prevention (ISBN: 2-912143-47-0; e-ISBN: 2912143764); *Eds. C. Petit, I. Al-Qadi and A. Millien*

PRO 38: 3rd International RILEM Workshop on Testing and Modelling the Chloride Ingress into Concrete (ISBN: 2-912143-48-9; e-ISBN: 2912143578); *Eds. C. Andrade and J. Kropp*

PRO 39: 6th International RILEM Symposium on Fibre-Reinforced Concretes—BEFIB 2004 (ISBN: 2-912143-51-9; e-ISBN: 2912143748); *Eds. M. Di Prisco, R. Felicetti and G. A. Plizzari*

PRO 40: International RILEM Conference on the Use of Recycled Materials in Buildings and Structures (ISBN: 2-912143-52-7; e-ISBN: 2912143756); *Eds. E. Vázquez, Ch. F. Hendriks and G. M. T. Janssen*

PRO 41: RILEM International Symposium on Environment-Conscious Materials and Systems for Sustainable Development (ISBN: 2-912143-55-1; e-ISBN: 2912143640); *Eds. N. Kashino and Y. Ohama*

PRO 42: SCC'2005—China: 1st International Symposium on Design, Performance and Use of Self-Consolidating Concrete (ISBN: 2-912143-61-6; e-ISBN: 2912143624); *Eds. Zhiwu Yu, Caijun Shi, Kamal Henri Khayat and Youjun Xie*

PRO 43: International RILEM Workshop on Bonded Concrete Overlays (e-ISBN: 2-912143-83-7); *Eds. J. L. Granju and J. Silfwerbrand*

PRO 44: 2nd International RILEM Workshop on Microbial Impacts on Building Materials (CD11) (e-ISBN: 2-912143-84-5); *Ed. M. Ribas Silva*

PRO 45: 2nd International Symposium on Nanotechnology in Construction, Bilbao (ISBN: 2-912143-87-X; e-ISBN: 2912143888); *Eds. Peter J. M. Bartos, Yolanda de Miguel and Antonio Porro*

PRO 46: Concrete Life'06—International RILEM-JCI Seminar on Concrete Durability and Service Life Planning: Curing, Crack Control, Performance in Harsh Environments (ISBN: 2-912143-89-6; e-ISBN: 291214390X); *Ed. K. Kovler*

PRO 47: International RILEM Workshop on Performance Based Evaluation and Indicators for Concrete Durability (ISBN: 978-2-912143-95-2; e-ISBN: 9782912143969); *Eds. V. Baroghel-Bouny, C. Andrade, R. Torrent and K. Scrivener*

PRO 48: 1st International RILEM Symposium on Advances in Concrete through Science and Engineering (e-ISBN: 2-912143-92-6); *Eds. J. Weiss, K. Kovler, J. Marchand, and S. Mindess*

PRO 49: International RILEM Workshop on High Performance Fiber Reinforced Cementitious Composites in Structural Applications (ISBN: 2-912143-93-4; e-ISBN: 2912143942); *Eds. G. Fischer and V. C. Li*

PRO 50: 1st International RILEM Symposium on Textile Reinforced Concrete (ISBN: 2-912143-97-7; e-ISBN: 2351580087); *Eds. Josef Hegger, Wolfgang Brameshuber and Norbert Will*

PRO 51: 2nd International Symposium on Advances in Concrete through Science and Engineering (ISBN: 2-35158-003-6; e-ISBN: 2-35158-002-8); *Eds. J. Marchand, B. Bissonnette, R. Gagné, M. Jolin and F. Paradis*

PRO 52: Volume Changes of Hardening Concrete: Testing and Mitigation (ISBN: 2-35158-004-4; e-ISBN: 2-35158-005-2); *Eds. O. M. Jensen, P. Lura and K. Kovler*

PRO 53: High Performance Fiber Reinforced Cement Composites—HPFRCC5 (ISBN: 978-2-35158-046-2; e-ISBN: 978-2-35158-089-9); *Eds. H. W. Reinhardt and A. E. Naaman*

PRO 54: 5th International RILEM Symposium on Self-Compacting Concrete (ISBN: 978-2-35158-047-9; e-ISBN: 978-2-35158-088-2); *Eds. G. De Schutter and V. Boel*

PRO 55: International RILEM Symposium Photocatalysis, Environment and Construction Materials (ISBN: 978-2-35158-056-1; e-ISBN: 978-2-35158-057-8); *Eds. P. Baglioni and L. Cassar*

PRO 56: International RILEM Workshop on Integral Service Life Modelling of Concrete Structures (ISBN 978-2-35158-058-5; e-ISBN: 978-2-35158-090-5); *Eds. R. M. Ferreira, J. Gulikers and C. Andrade*

PRO 57: RILEM Workshop on Performance of cement-based materials in aggressive aqueous environments (e-ISBN: 978-2-35158-059-2); *Ed. N. De Belie*

PRO 58: International RILEM Symposium on Concrete Modelling—CONMOD'08 (ISBN: 978-2-35158-060-8; e-ISBN: 978-2-35158-076-9); *Eds. E. Schlangen and G. De Schutter*

PRO 59: International RILEM Conference on On Site Assessment of Concrete, Masonry and Timber Structures—SACoMaTiS 2008 (ISBN set: 978-2-35158-061-5; e-ISBN: 978-2-35158-075-2); *Eds. L. Binda, M. di Prisco and R. Felicetti*

PRO 60: Seventh RILEM International Symposium on Fibre Reinforced Concrete: Design and Applications—BEFIB 2008 (ISBN: 978-2-35158-064-6; e-ISBN: 978-2-35158-086-8); *Ed. R. Gettu*

PRO 61: 1st International Conference on Microstructure Related Durability of Cementitious Composites 2 vol., (ISBN: 978-2-35158-065-3; e-ISBN: 978-2-35158-084-4); *Eds. W. Sun, K. van Breugel, C. Miao, G. Ye and H. Chen*

PRO 62: NSF/ RILEM Workshop: In-situ Evaluation of Historic Wood and Masonry Structures (e-ISBN: 978-2-35158-068-4); *Eds. B. Kasal, R. Anthony and M. Drdácký*

PRO 63: Concrete in Aggressive Aqueous Environments: Performance, Testing and Modelling, 2 vol., (ISBN: 978-2-35158-071-4; e-ISBN: 978-2-35158-082-0); *Eds. M. G. Alexander and A. Bertron*

PRO 64: Long Term Performance of Cementitious Barriers and Reinforced Concrete in Nuclear Power Plants and Waste Management—NUCPERF 2009 (ISBN: 978-2-35158-072-1; e-ISBN: 978-2-35158-087-5); *Eds. V. L'Hostis, R. Gens and C. Gallé*

PRO 65: Design Performance and Use of Self-consolidating Concrete—SCC'2009 (ISBN: 978-2-35158-073-8; e-ISBN: 978-2-35158-093-6); *Eds. C. Shi, Z. Yu, K. H. Khayat and P. Yan*

PRO 66: 2nd International RILEM Workshop on Concrete Durability and Service Life Planning—ConcreteLife'09 (ISBN: 978-2-35158-074-5; ISBN: 978-2-35158-074-5); *Ed. K. Kovler*

PRO 67: Repairs Mortars for Historic Masonry (e-ISBN: 978-2-35158-083-7); *Ed. C. Groot*

PRO 68: Proceedings of the 3rd International RILEM Symposium on 'Rheology of Cement Suspensions such as Fresh Concrete (ISBN 978-2-35158-091-2; e-ISBN: 978-2-35158-092-9); *Eds. O. H. Wallevik, S. Kubens and S. Oesterheld*

PRO 69: 3rd International PhD Student Workshop on 'Modelling the Durability of Reinforced Concrete (ISBN: 978-2-35158-095-0); *Eds. R. M. Ferreira, J. Gulikers and C. Andrade*

PRO 70: 2nd International Conference on 'Service Life Design for Infrastructure' (ISBN set: 978-2-35158-096-7, e-ISBN: 978-2-35158-097-4); *Eds. K. van Breugel, G. Ye and Y. Yuan*

PRO 71: Advances in Civil Engineering Materials—The 50-year Teaching Anniversary of Prof. Sun Wei' (ISBN: 978-2-35158-098-1; e-ISBN: 978-2-35158-099-8); *Eds. C. Miao, G. Ye and H. Chen*

PRO 72: First International Conference on 'Advances in Chemically-Activated Materials—CAM'2010' (2010), 264 pp., ISBN: 978-2-35158-101-8; e-ISBN: 978-2-35158-115-5; *Eds. Caijun Shi and Xiaodong Shen*

PRO 73: 2nd International Conference on 'Waste Engineering and Management—ICWEM 2010' (2010), 894 pp., ISBN: 978-2-35158-102-5; e-ISBN: 978-2-35158-103-2, *Eds. J. Zh. Xiao, Y. Zhang, M. S. Cheung and R. Chu*

PRO 74: International RILEM Conference on 'Use of Superabsorsorbent Polymers and Other New Adddtitives in Concrete' (2010) 374 pp., ISBN: 978-2-35158-104-9; e-ISBN: 978-2-35158-105-6; *Eds. O.M. Jensen, M.T. Hasholt, and S. Laustsen*

PRO 75: International Conference on 'Material Science—2nd ICTRC—Textile Reinforced Concrete—Theme 1' (2010) 436 pp., ISBN: 978-2-35158-106-3; e-ISBN: 978-2-35158-107-0; *Ed. W. Brameshuber*

PRO 76: International Conference on 'Material Science—HetMat—Modelling of Heterogeneous Materials—Theme 2' (2010) 255 pp., ISBN: 978-2-35158-108-7; e-ISBN: 978-2-35158-109-4; *Ed. W. Brameshuber*

PRO 77: International Conference on 'Material Science—AdIPoC—Additions Improving Properties of Concrete—Theme 3' (2010) 459 pp., ISBN: 978-2-35158-110-0; e-ISBN: 978-2-35158-111-7; *Ed. W. Brameshuber*

PRO 78: 2nd Historic Mortars Conference and RILEM TC 203-RHM Final Workshop—HMC2010 (2010) 1416 pp., e-ISBN: 978-2-35158-112-4; *Eds. J. Válek, C. Groot and J. J. Hughes*

PRO 79: International RILEM Conference on Advances in Construction Materials Through Science and Engineering (2011) 213 pp., ISBN: 978-2-35158-116-2, e-ISBN: 978-2-35158-117-9; *Eds. Christopher Leung and K.T. Wan*

PRO 80: 2nd International RILEM Conference on Concrete Spalling due to Fire Exposure (2011) 453 pp., ISBN: 978-2-35158-118-6; e-ISBN: 978-2-35158-119-3; *Eds. E.A.B. Koenders and F. Dehn*

PRO 81: 2nd International RILEM Conference on Strain Hardening Cementitious Composites (SHCC2-Rio) (2011) 451 pp., ISBN: 978-2-35158-120-9; e-ISBN: 978-2-35158-121-6; *Eds. R.D. Toledo Filho, F.A. Silva, E.A.B. Koenders and E.M. R. Fairbairn*

PRO 82: 2nd International RILEM Conference on Progress of Recycling in the Built Environment (2011) 507 pp., e-ISBN: 978-2-35158-122-3; *Eds. V.M. John, E. Vazquez, S.C. Angulo and C. Ulsen*

PRO 83: 2nd International Conference on Microstructural-related Durability of Cementitious Composites (2012) 250 pp., ISBN: 978-2-35158-129-2; e-ISBN: 978-2-35158-123-0; *Eds. G. Ye, K. van Breugel, W. Sun and C. Miao*

PRO 84: CONSEC13—Seventh International Conference on Concrete under Severe Conditions—Environment and Loading (2013) 1930 pp., ISBN: 978-2-35158-124-7; e-ISBN: 978-2- 35158-134-6; *Eds. Z.J. Li, W. Sun, C.W. Miao, K. Sakai, O.E. Gjorv and N. Banthia*

PRO 85: RILEM-JCI International Workshop on Crack Control of Mass Concrete and Related issues concerning Early-Age of Concrete Structures—ConCrack 3—Control of Cracking in Concrete Structures 3 (2012) 237 pp., ISBN: 978-2-35158-125-4; e-ISBN: 978-2-35158-126-1; *Eds. F. Toutlemonde and J.-M. Torrenti*

PRO 86: International Symposium on Life Cycle Assessment and Construction (2012) 414 pp., ISBN: 978-2-35158-127-8, e-ISBN: 978-2-35158-128-5; *Eds. A. Ventura and C. de la Roche*

PRO 87: UHPFRC 2013—RILEM-fib-AFGC International Symposium on Ultra-High Performance Fibre-Reinforced Concrete (2013), ISBN: 978-2-35158-130-8, e-ISBN: 978-2-35158-131-5; *Eds. F. Toutlemonde*

PRO 88: 8th RILEM International Symposium on Fibre Reinforced Concrete (2012) 344 pp., ISBN: 978-2-35158-132-2; e-ISBN: 978-2-35158-133-9; *Eds. Joaquim A.O. Barros*

PRO 89: RILEM International workshop on performance-based specification and control of concrete durability (2014) 678 pp., ISBN: 978-2-35158-135-3; e-ISBN: 978-2-35158-136-0; *Eds. D. Bjegović, H. Beushausen and M. Serdar*

PRO 90: 7th RILEM International Conference on Self-Compacting Concrete and of the 1st RILEM International Conference on Rheology and Processing of Construction Materials (2013) 396 pp., ISBN: 978-2-35158-137-7; e-ISBN: 978-2-35158-138-4; *Eds. Nicolas Roussel and Hela Bessaies-Bey*

PRO 91: CONMOD 2014—RILEM International Symposium on Concrete Modelling (2014), ISBN: 978-2-35158-139-1; e-ISBN: 978-2-35158-140-7; *Eds. Kefei Li, Peiyu Yan and Rongwei Yang*

PRO 92: CAM 2014—2nd International Conference on advances in chemically-activated materials (2014) 392 pp., ISBN: 978-2-35158-141-4; e-ISBN: 978-2-35158-142-1; *Eds. Caijun Shi and Xiadong Shen*

PRO 93: SCC 2014—3rd International Symposium on Design, Performance and Use of Self-Consolidating Concrete (2014) 438 pp., ISBN: 978-2-35158-143-8; e-ISBN: 978-2-35158-144-5; *Eds. Caijun Shi, Zhihua Ou and Kamal H. Khayat*

PRO 94 (online version): HPFRCC-7—7th RILEM conference on High performance fiber reinforced cement composites (2015), e-ISBN: 978-2-35158-146-9; *Eds. H.W. Reinhardt, G.J. Parra-Montesinos and H. Garrecht*

PRO 95: International RILEM Conference on Application of superabsorbent polymers and other new admixtures in concrete construction (2014), ISBN: 978-2-35158-147-6; e-ISBN: 978-2-35158-148-3; *Eds. Viktor Mechtcherine and Christof Schroefl*

PRO 96 (online version): XIII DBMC: XIII International Conference on Durability of Building Materials and Components (2015), e-ISBN: 978-2-35158-149-0; *Eds. M. Quattrone and V.M. John*

PRO 97: SHCC3—3rd International RILEM Conference on Strain Hardening Cementitious Composites (2014), ISBN: 978-2-35158-150-6; e-ISBN: 978-2-35158-151-3; *Eds. E. Schlangen, M.G. Sierra Beltran, M. Lukovic and G. Ye*

PRO 98: FERRO-11—11th International Symposium on Ferrocement and 3rd ICTRC—International Conference on Textile Reinforced Concrete (2015), ISBN: 978-2-35158-152-0; e-ISBN: 978-2-35158-153-7; *Ed. W. Brameshuber*

PRO 99 (online version): ICBBM 2015—1st International Conference on Bio-Based Building Materials (2015), e-ISBN: 978-2-35158-154-4; *Eds. S. Amziane and M. Sonebi*

PRO 100: SCC16—RILEM Self-Consolidating Concrete Conference (2016), ISBN: 978-2-35158-156-8; e-ISBN: 978-2-35158-157-5; *Ed. Kamal H. Kayat*

PRO 101 (online version): III Progress of Recycling in the Built Environment (2015), e-ISBN: 978-2-35158-158-2; *Eds I. Martins, C. Ulsen and S. C. Angulo*

PRO 102 (online version): RILEM Conference on Microorganisms-Cementitious Materials Interactions (2016), e-ISBN: 978-2-35158-160-5; *Eds. Alexandra Bertron, Henk Jonkers and Virginie Wiktor*

PRO 103 (online version): ACESC'16—Advances in Civil Engineering and Sustainable Construction (2016), e-ISBN: 978-2-35158-161-2; *Eds. T.Ch. Madhavi, G. Prabhakar, Santhosh Ram and P.M. Rameshwaran*

PRO 104 (online version): SSCS'2015—Numerical Modeling—Strategies for Sustainable Concrete Structures (2015), e-ISBN: 978-2-35158-162-9

PRO 105: 1st International Conference on UHPC Materials and Structures (2016), ISBN: 978-2-35158-164-3; e-ISBN: 978-2-35158-165-0

PRO 106: AFGC-ACI-fib-RILEM International Conference on Ultra-High-Performance Fibre-Reinforced Concrete—UHPFRC 2017 (2017), ISBN: 978-2-35158-166-7; e-ISBN: 978-2-35158-167-4; *Eds. François Toutlemonde and Jacques Resplendino*

PRO 107 (online version): XIV DBMC—14th International Conference on Durability of Building Materials and Components (2017), e-ISBN: 978-2-35158-159-9; *Eds. Geert De Schutter, Nele De Belie, Arnold Janssens and Nathan Van Den Bossche*

PRO 108: MSSCE 2016—Innovation of Teaching in Materials and Structures (2016), ISBN: 978-2-35158-178-0; e-ISBN: 978-2-35158-179-7; *Ed. Per Goltermann*

PRO 109 (2 volumes): MSSCE 2016—Service Life of Cement-Based Materials and Structures (2016), ISBN Vol. 1: 978-2-35158-170-4; Vol. 2: 978-2-35158-171-4; Set Vol. 1&2: 978-2-35158-172-8; e-ISBN : 978-2-35158-173-5; *Eds. Miguel Azenha, Ivan Gabrijel, Dirk Schlicke, Terje Kanstad and Ole Mejlhede Jensen*

PRO 110: MSSCE 2016—Historical Masonry (2016), ISBN: 978-2-35158-178-0; e-ISBN: 978-2-35158-179-7; *Eds. Inge Rörig-Dalgaard and Ioannis Ioannou*

PRO 111: MSSCE 2016—Electrochemistry in Civil Engineering (2016); ISBN: 978-2-35158-176-6; e-ISBN: 978-2-35158-177-3; *Ed. Lisbeth M. Ottosen*

PRO 112: MSSCE 2016—Moisture in Materials and Structures (2016), ISBN: 978-2-35158-178-0; e-ISBN: 978-2-35158-179-7; *Eds. Kurt Kielsgaard Hansen, Carsten Rode and Lars-Olof Nilsson*

PRO 113: MSSCE 2016—Concrete with Supplementary Cementitious Materials (2016), ISBN: 978-2-35158-178-0; e-ISBN: 978-2-35158-179-7; *Eds. Ole Mejlhede Jensen, Konstantin Kovler and Nele De Belie*

PRO 114: MSSCE 2016—Frost Action in Concrete (2016), ISBN: 978-2-35158-182-7; e-ISBN: 978-2-35158-183-4; *Eds. Marianne Tange Hasholt, Katja Fridh and R. Doug Hooton*

PRO 115: MSSCE 2016—Fresh Concrete (2016), ISBN: 978-2-35158-184-1; e-ISBN: 978-2-35158-185-8; *Eds. Lars N. Thrane, Claus Pade, Oldrich Svec and Nicolas Roussel*

PRO 116: BEFIB 2016—9th RILEM International Symposium on Fiber Reinforced Concrete (2016), ISBN: 978-2-35158-187-2; e-ISBN: 978-2-35158-186-5; *Eds. N. Banthia, M. di Prisco and S. Soleimani-Dashtaki*

PRO 117: 3rd International RILEM Conference on Microstructure Related Durability of Cementitious Composites (2016), ISBN: 978-2-35158-188-9; e-ISBN: 978-2-35158-189-6; *Eds. Changwen Miao, Wei Sun, Jiaping Liu, Huisu Chen, Guang Ye and Klaas van Breugel*

PRO 118 (4 volumes): International Conference on Advances in Construction Materials and Systems (2017), ISBN Set: 978-2-35158-190-2; Vol. 1: 978-2-35158-193-3; Vol. 2: 978-2-35158-194-0; Vol. 3: ISBN:978-2-35158-195-7; Vol. 4: ISBN:978-2-35158-196-4; e-ISBN: 978-2-35158-191-9; *Ed. Manu Santhanam*

PRO 119 (online version): ICBBM 2017—Second International RILEM Conference on Bio-based Building Materials, (2017), e-ISBN: 978-2-35158-192-6; *Ed. Sofiane Amziane*

PRO 120 (2 volumes): EAC-02—2nd International RILEM/COST Conference on Early Age Cracking and Serviceability in Cement-based Materials and Structures, (2017), Vol. 1: 978-2-35158-199-5, Vol. 2: 978-2-35158-200-8, Set: 978-2-35158-197-1, e-ISBN: 978-2-35158-198-8; *Eds. Stéphanie Staquet and Dimitrios Aggelis*

PRO 121 (2 volumes): SynerCrete18: Interdisciplinary Approaches for Cementbased Materials and Structural Concrete: Synergizing Expertise and Bridging Scales of Space and Time, (2018), Set: 978-2-35158-202-2, Vol.1: 978-2-35158-211-4, Vol.2: 978-2-35158-212-1, e-ISBN: 978-2-35158-203-9; *Eds. Miguel Azenha, Dirk Schlicke, Farid Benboudjema, Agnieszka Knoppik*

PRO 122: SCC'2018 China—Fourth International Symposium on Design, Performance and Use of Self-Consolidating Concrete, (2018), ISBN: 978-2-35158-204-6, e-ISBN: 978-2-35158-205-3; *Eds. C. Shi, Z. Zhang, K. H. Khayat*

PRO 123: Final Conference of RILEM TC 253-MCI: Microorganisms-Cementitious Materials Interactions (2018), Set: 978-2-35158-207-7, Vol.1: 978-2-35158-209-1, Vol.2: 978-2-35158-210-7, e-ISBN: 978-2-35158-206-0; *Ed. Alexandra Bertron*

PRO 124 (online version): Fourth International Conference Progress of Recycling in the Built Environment (2018), e-ISBN: 978-2-35158-208-4; *Eds. Isabel M. Martins, Carina Ulsen, Yury Villagran*

PRO 125 (online version): SLD4—4th International Conference on Service Life Design for Infrastructures (2018), e-ISBN: 978-2-35158-213-8; *Eds. Guang Ye, Yong Yuan, Claudia Romero Rodriguez, Hongzhi Zhang, Branko Savija*

PRO 126: Workshop on Concrete Modelling and Material Behaviour in honor of Professor Klaas van Breugel (2018), ISBN: 978-2-35158-214-5, e-ISBN: 978-2-35158-215-2; *Ed. Guang Ye*

PRO 127 (online version): CONMOD2018—Symposium on Concrete Modelling (2018), e-ISBN: 978-2-35158-216-9; *Eds. Erik Schlangen, Geert de Schutter, Branko Savija, Hongzhi Zhang, Claudia Romero Rodriguez*

PRO 128: SMSS2019—International Conference on Sustainable Materials, Systems and Structures (2019), ISBN: 978-2-35158-217-6, e-ISBN: 978-2-35158-218-3

PRO 129: 2nd International Conference on UHPC Materials and Structures (UHPC2018-China), ISBN: 978-2-35158-219-0, e-ISBN: 978-2-35158-220-6

PRO 130: 5th Historic Mortars Conference (2019), ISBN: 978-2-35158-221-3, e-ISBN: 978-2-35158-222-0; *Eds. José Ignacio Álvarez, José María Fernández, Íñigo Navarro, Adrián Durán, Rafael Sirera*

PRO 131 (online version): 3rd International Conference on Bio-Based Building Materials (ICBBM2019), e-ISBN: 978-2-35158-229-9; *Eds. Mohammed Sonebi, Sofiane Amziane, Jonathan Page*

PRO 132: IRWRMC'18—International RILEM Workshop on Rheological Measurements of Cement-based Materials (2018), ISBN: 978-2-35158-230-5, e-ISBN: 978-2-35158-231-2; *Eds. Chafika Djelal, Yannick Vanhove*

PRO 133 (online version): CO2STO2019—International Workshop CO2 Storage in Concrete (2019), e-ISBN: 978-2-35158-232-9; *Eds. Assia Djerbi, Othman Omikrine-Metalssi, Teddy Fen-Chong*

RILEM Reports (REP)

Report 19: Considerations for Use in Managing the Aging of Nuclear Power Plant Concrete Structures (ISBN: 2-912143-07-1); *Ed. D. J. Naus*

Report 20: Engineering and Transport Properties of the Interfacial Transition Zone in Cementitious Composites (ISBN: 2-912143-08-X); *Eds. M. G. Alexander, G. Arliguie, G. Ballivy, A. Bentur and J. Marchand*

Report 21: Durability of Building Sealants (ISBN: 2-912143-12-8); *Ed. A. T. Wolf*

Report 22: Sustainable Raw Materials—Construction and Demolition Waste (ISBN: 2-912143-17-9); *Eds. C. F. Hendriks and H. S. Pietersen*

Report 23: Self-Compacting Concrete state-of-the-art report (ISBN: 2-912143-23-3); *Eds. Å. Skarendahl and Ö. Petersson*

Report 24: Workability and Rheology of Fresh Concrete: Compendium of Tests (ISBN: 2-912143-32-2); *Eds. P. J. M. Bartos, M. Sonebi and A. K. Tamimi*

Report 25: Early Age Cracking in Cementitious Systems (ISBN: 2-912143-33-0); *Ed. A. Bentur*

Report 26: Towards Sustainable Roofing (Joint Committee CIB/RILEM) (CD 07) (e-ISBN 978-2-912143-65-5); *Eds. Thomas W. Hutchinson and Keith Roberts*

Report 27: Condition Assessment of Roofs (Joint Committee CIB/RILEM) (CD 08) (e-ISBN 978-2-912143-66-2); *Ed. CIB W 83/RILEM TC166-RMS*

Report 28: Final report of RILEM TC 167-COM 'Characterisation of Old Mortars with Respect to Their Repair (ISBN: 978-2-912143-56-3); *Eds. C. Groot, G. Ashall and J. Hughes*

Report 29: Pavement Performance Prediction and Evaluation (PPPE): Interlaboratory Tests (e-ISBN: 2-912143-68-3); *Eds. M. Partl and H. Piber*

Report 30: Final Report of RILEM TC 198-URM 'Use of Recycled Materials' (ISBN: 2-912143-82-9; e-ISBN: 2-912143-69-1); *Eds. Ch. F. Hendriks, G. M. T. Janssen and E. Vázquez*

Report 31: Final Report of RILEM TC 185-ATC 'Advanced testing of cement-based materials during setting and hardening' (ISBN: 2-912143-81-0; e-ISBN: 2-912143-70-5); *Eds. H. W. Reinhardt and C. U. Grosse*

Report 32: Probabilistic Assessment of Existing Structures. A JCSS publication (ISBN 2-912143-24-1); *Ed. D. Diamantidis*

Report 33: State-of-the-Art Report of RILEM Technical Committee TC 184-IFE 'Industrial Floors' (ISBN 2-35158-006-0); *Ed. P. Seidler*

Report 34: Report of RILEM Technical Committee TC 147-FMB 'Fracture mechanics applications to anchorage and bond' Tension of Reinforced Concrete Prisms—Round Robin Analysis and Tests on Bond (e-ISBN 2-912143-91-8); *Eds. L. Elfgren and K. Noghabai*

Report 35: Final Report of RILEM Technical Committee TC 188-CSC 'Casting of Self Compacting Concrete' (ISBN 2-35158-001-X; e-ISBN: 2-912143-98-5); *Eds. Å. Skarendahl and P. Billberg*

Report 36: State-of-the-Art Report of RILEM Technical Committee TC 201-TRC 'Textile Reinforced Concrete' (ISBN 2-912143-99-3); *Ed. W. Brameshuber*

Report 37: State-of-the-Art Report of RILEM Technical Committee TC 192-ECM 'Environment-conscious construction materials and systems' (ISBN: 978-2-35158-053-0); *Eds. N. Kashino, D. Van Gemert and K. Imamoto*

Report 38: State-of-the-Art Report of RILEM Technical Committee TC 205-DSC 'Durability of Self-Compacting Concrete' (ISBN: 978-2-35158-048-6); *Eds. G. De Schutter and K. Audenaert*

Report 39: Final Report of RILEM Technical Committee TC 187-SOC 'Experimental determination of the stress-crack opening curve for concrete in tension' (ISBN 978-2-35158-049-3); *Ed. J. Planas*

Report 40: State-of-the-Art Report of RILEM Technical Committee TC 189-NEC 'Non-Destructive Evaluation of the Penetrability and Thickness of the Concrete Cover' (ISBN 978-2-35158-054-7); *Eds. R. Torrent and L. Fernández Luco*

Report 41: State-of-the-Art Report of RILEM Technical Committee TC 196-ICC 'Internal Curing of Concrete' (ISBN 978-2-35158-009-7); *Eds. K. Kovler and O. M. Jensen*

Report 42: 'Acoustic Emission and Related Non-destructive Evaluation Techniques for Crack Detection and Damage Evaluation in Concrete'—Final Report of RILEM Technical Committee 212-ACD (e-ISBN: 978-2-35158-100-1); *Ed. M. Ohtsu*

Report 45: Repair Mortars for Historic Masonry—State-of-the-Art Report of RILEM Technical Committee TC 203-RHM (e-ISBN: 978-2-35158-163-6); *Eds. Paul Maurenbrecher and Caspar Groot*

Report 46: Surface delamination of concrete industrial ffioors and other durability related aspects guide—Report of RILEM Technical Committee TC 268-SIF (e-ISBN: 978-2-35158-201-5); *Ed. Valerie Pollet*

Contents

Reinforcement of Timber Elements in Existing Structures

Thomas Tannert, Philipp Dietsch, and Jorge Branco

Abstract Wood and engineered wood products, herein also referred to as timber, have been used as structural building material for centuries and countless examples demonstrate its longevity if properly designed, built, maintained and assessed [1, 2]. The more recent development of new engineered wood products, connector systems and growing awareness about sustainability in the construction sector have led to legislative changes in the building sector and as a consequence also a significant widening in the range of structural applications of timber [3, 4].

In parallel, there is also a growing need for the maintenance and upgrading of existing buildings for economic, environmental, historical and social concerns, such as the need for more affordable housing. Worldwide, a large proportion of the existing building stock is more than 50 years old; many of these buildings need to be adapted for present and future requirements. For example, about 50% of all construction in Europe is already related to existing buildings [5, 6]. The European Commission mandates the development of rules for the assessment of existing structures and their reinforcement [7]. The need for structural reinforcement of timber buildings may become necessary from motivations such as change of use, changes in regulatory specifications, interventions to increase seismic resistance, deterioration due to poor maintenance, or exceptional damaging incidents [8].

The structural reinforcement of timber structures can be grouped into two main categories: (i) addition of new systems to support the existing structure; and (ii) incorporation of reinforcements into existing elements and joints. In the first category, the options include adding new lateral and gravity load resisting systems such as braces, shear walls or timber-steel and timber-concrete composite floors. In the

T. Tannert
University of Northern British Columbia, Prince George, Canada

P. Dietsch
Unit of Timber Engineering, Department of Structural Engineering and Material Sciences, University of Innsbruck, Innsbruck, Austria

J. Branco (✉)
Department of Civil Engineering, ISISE, University of Minho, Guimarães, Portugal
e-mail: jbranco@civil.uminho.pt

© RILEM 2021
J. Branco et al. (eds.), *Reinforcement of Timber Elements in Existing Structures*,
RILEM State-of-the-Art Reports 33,
https://doi.org/10.1007/978-3-030-67794-7_1

1

second category, the options range from mechanical fasteners like glued-in rods (GiR) and self-tapping screws (STS), adhesive systems in combination fibre-reinforced polymers (FRP), and most recently nanotechnology (e.g. carbon nanotubes with polymeric resins). Despite the current lack of design approaches in international standards, reinforcement of structural systems such as light-frame wood lateral load resisting systems, timber floors in existing buildings and elements and joints exposed to stresses perpendicular to the grain can be considered state-of-the art in timber engineering practice.

In this state of the art report, an overview of the existing techniques for the reinforcement of timber elements, joints and structures is provided. The report consists of two parts. In Part I, the state of the art on reinforcement techniques and the current status of standardization is summarized. Specifically, STS, GiR, FRP, and nanotechnology are covered. In Part II, several applications of reinforcement are discussed. These include traditional structures, traditional timber frame walls, light-frame shear walls, roofs, floors, and carpentry joints. In the following a short summary is provided for each chapter.

In Chapter "Self-tapping Screws as Reinforcement for Structural Timber Elements", STS are presented. STS are made of hardened steel and exhibit high yield, tensile, and torsional strengths as well as high withdrawal resistance. The thread provides a continuous mechanical connection along the embedded length, which makes STS efficient for reinforcing timber elements and connections prone to splitting. Their high axial stiffness and load-carrying capacity in combination with easy handling make them one of the most popular and also economical choices as reinforcing element. The material properties of STS and screwed-in threaded rods, mechanical models for timber members reinforced with STS as well as the behavior of reinforced timber members under shrinkage is discussed. The STS should be designed to carry the full stresses that can lead to brittle failure modes. The reinforcing effect of STS is described as reliable in cases where they are installed in timber members featuring a wood moisture content close to the equilibrium moisture content of the timber member in use.

In Chapter "Glued-in Rods as Reinforcement for Timber Structural Elements", GiR are presented. GiR are high strength connectors that are concealed inside the wood member which is both an architecturally pleasing feature and provides the joint with excellent fire protection. GiR are composed of multiple components, namely rod (mostly steel), timber and adhesive. GiR as reinforcement are highly complex systems of different materials which require adequate mechanical models for the determination of stress distribution and load-carrying capacity. The complex stress distribution along the bond line between rod and wood is influenced—amongst other factors—by the adhesive properties. Especially the failure behaviour of the adhesive has an important impact on the failure behaviour of GiR. Models for the determination of the shear stress distribution in the bondline along the rod are discussed and influences on the stress distribution and strength of GiR are evaluated. Aspects on how to increase the load-carrying capacity and how to reach a ductile failure behaviour are presented.

In Chapter "Fiber-Reinforced Polymers as Reinforcement for Timber Structural Elements", the recent development of FRP reinforced timber elements and joints is discussed. The constituents of FRP composite materials, namely the fibres and the matrix, are introduced and their typical applications in the reinforcement of timber elements and joints are discussed. The successful implementation of FRP and wood bonding are highly dependent on various factors, from initial raw materials properties to proper surface preparation when implementing the intervention. For each step, the process for application of bonding should follow strictly the specifications to achieve a good bonding quality. The long-term performance of adhesively bonded FRP reinforcements under environmental effects (i.e. moisture, temperature, coupled moisture and temperature, and fire) are reviewed. Furthermore, the in situ quality control for bonding FRP and wood substrate and assessment of FRP-reinforced wood structures are described. Finally, future perspectives on the application of such hybrid systems are given. Based on these discussions, it is made clear that using FRP with appropriate adhesive bonding can effectively improve the stiffness and strength of wood elements.

In Chapter "Nanocomposites as Reinforcement for Timber Structural Elements", an overview of technological innovations in nanocomposite materials for protection and reinforcement of timber is presented, starting from the definition of nanotechnologies applied in the field of construction field and architectural heritage. The role of different nano-coatings, their wood surface protection functions and compatibility with the different wood species are reviewed. It is also shown, that wood as a natural nanocomposite, can potentially offer important applications in the field of nanotechnologies. In addition to the required performance measures such as durability and weather resistance, good adhesion to the substrate, transparency, sustainability for the production process, nano-structured coatings can also introduce additional surface functionality such as self-cleaning, fire- and scratch resistance and antibacterial properties. Research projects with a special focus on reinforcement of historic timber joints are also presented in this chapter.

In Chapter "Reinforcement of Timber Structures: Standardization Towards a New Section for EC 5", a link between techniques and applications is provided though a discussion of the current state of the art of standardization with respect to the reinforcement of timber structures. While new materials and methods for reinforcement have been developed and are now used in practice, the design standards in their current editions lack specific guidance to design reinforcements for timber members and joints. To close this gap in the new generation of Eurocode 5, CEN/TC 250/SC 5, the standardization committee responsible for drafting the European Timber Design standard, established a Working Group 7 "Reinforcement" on this item. This chapter presents the approach to this task, the work items, the work plan, the structure as well as design approaches and related background information of the proposed Eurocode 5 section.

In Chapter "Seismic Reinforcement of Traditional Timber Structures", the reinforcement of traditional buildings, particularly to meet seismic demands, is discussed. Earthquakes cause damage and destroy a sizeable number of buildings across the globe. Yet timber buildings are often considered earthquake-resistant, provided that load-path continuity is provided, the joints are intact, and moisture-induced problems are kept at bay. However, the high costs and difficulties involved in executing interventions needed to meet safety requirements prescribed by current building codes act as a deterrent. Therefore, it is important to identify the inherent seismic-resistant features as well as the deficiencies of traditional timber constructions and review the various strengthening, retrofitting and upgrading measures that have been developed to enhance the safety of such structures. The effectiveness of different strengthening techniques has been proven on the basis of results from experimental tests carried out on components of timber structures ranging from joints and beams to full-scale shear walls, roof trusses and floor slabs. The success stories and failures of past interventions implemented in practice also shown as they play an instrumental role in identifying effective and economical strengthening solutions for traditional timber structures.

In Chapter "Reinforcement of Traditional Timber Frame Walls", timber frame walls, which constitute an important cultural heritage of different parts of the world, are introduced and the necessity for interventions in for their preservation. An overview of the possible retrofitting techniques is presented, focusing on their advantages and disadvantages and their effect on the overall wall behaviour. Traditional interventions such as timber-to-timber, metal fasteners and steel elements are able to restore and improve the capacity of walls. As the response of the walls, particularly to horizontal actions, is governed by their connections, retrofitting is usually concentrated at the joints, but interventions can also be carried out on timber members or the infill. Though the focus has been on interventions for seismic actions, due to the adoption of such walls as shear walls, these interventions can also be applied for other objectives. Little research is available on the reinforcement of traditional timber frame walls, but the results available on walls and other elements have shown possible effective interventions.

In Chapter "Reinforcement of Light-Frame Wood Structures", the reinforcement of light-frame wood lateral structural systems is discussed. Post-earthquake evaluations of the performance of such buildings have pointed to a number of deficiencies that make them susceptible to high levels of damage and collapse during earthquakes. This is mostly caused by the fact that many of these buildings were designed and built before the adoption of modern seismic building codes. This chapter focuses firstly on defining the deficiencies that are common in light-frame wood structures, including: weak first storey; weak roof or floor diaphragms; shear walls with insufficient strength; inadequate load path; geometric and mass eccentricities; and brittle components. Subsequently, the chapter describes conventional as well as novel retrofit solutions that are available in specific codes and guidelines.

In Chapter "Reinforcement of Historic Timber Roofs", the reinforcement of historic timber roofs is presented, covering their characteristics, their advantages and disadvantages, to provide guidance for selecting and designing a solution. When new loads are introduced, reinforcement of traditional timber roofs is needed. If the decay of timber elements is too large, then local replacement of the decayed part is the best solution. When interventions are necessary, specific reliable on-site assessment techniques are required to determine the appropriate level of intervention. This point remains very important to evaluate the replacement, repair and reinforcing solutions along with the cultural significance of each case, the know-how and the associated project costs. Evaluation of the durability of the intervention work carried out with new innovative techniques is necessary too. The use of wood as reinforcement offers an interesting feature in conservation: compatibility. An overview of the main materials, and the techniques used for reinforcing roofs is presented in form of case studies.

In Chapter "Retrofitting of Traditional Timber Floors", timber floors—as a critical component of many historical and modern buildings—are discussed. Due to incorrect design and construction, effects of deterioration, change in use or functional requirements, floors frequently need to be retrofitted. This chapter reviews the typical reasons for retrofitting of existing timber floors and describes and critically evaluates a range of methods to improve their strength and stiffness, both in-plane and out-of-plane. The review shows that this is a very active area of research.

In Chapter "Reinforcement of Traditional Carpentry Joints", the state-of-the art on the reinforcement of traditional carpentry joints is presented by reviewing different standards, in situ case studies, as well as analytical, numerical and experimental research works. The geometrical typologies of traditional carpentry joints but also their failure mechanisms such as compressive crushing, shear and tensile cracks are introduced to better understand how they work within timber structures. Based on this knowledge, the question of reinforcement can be tackled. To this end, several reinforcement strategies are presented by defining their objectives, methodologies, traditional and contemporary techniques, performance criteria and applications. When assessing and reinforcing traditional carpentry joints, some challenges may come up, namely their design based on the ratios of stiffness and load-bearing capacities; these properties should thus be a focal point for further research in the near future.

Concluding it can be stated that reinforcement techniques and design procedures have already been clarified to an extent satisfying safety requirements and engineering needs are available for numerous applications of timber joints, elements and systems. While existing methods can be useful in extending the life of existing structures, it should be expected that new and better methods will be developed in the future. Future research and development should focus on the determination of stiffness properties of the reinforcement in the timber as well as a better understanding and quantification of the potentially harmful effect of reinforcement restricting the free shrinkage or swelling of the timber elements. Also, a better understanding of reinforcement of structural systems on the whole building's lateral performance is crucial. The continuous dissemination of the knowledge can provide the necessary tools for all members of a multi-disciplinary team to evaluate the existing condition of a timber structure and moreover to select the proper interventions that will preserve the project's authenticity. When timber structures are reliably maintained and—if needed—their joints, elements and systems are repaired or reinforced, structural failures and unnecessary replacements can be avoided.

References

1. Kasal B, Tannert T (eds) (2011) In situ assessment of structural timber. RILEM state-of-the-art reports, vol 7, 129 p
2. Dietsch P, Köhler J (eds) (2010) Assessment of timber structures, Shaker. ISBN 978-3-8322-9513-4
3. Green M, Karsh JE (2012) Tall wood-the case for tall wood buildings. Wood Enterprise Coalition, Vancouver, Canada
4. Dietsch P, Tannert T (2015) Assessing the integrity of glued-laminated timber elements. Const. Build. Mater. 101(2):1259–1270
5. Itard L, Meijer F, Evert V, Hoiting H (2008) Building renovation and modernisation in Europe: state of the art review. OTB Research Institute for Housing, Delft
6. Housing Statistics in the European Union (2004) National board of housing, building and planning. Boverket, ISBN 91-7147-865-5, 2005
7. EN/TC 50—N848 (2010) Programming mandate addressed to CEN in the field of she structural Eurocodes (M/466 EN)
8. Dietsch P, Harte A (eds) (2015) Reinforcement of timber structures: a state-of-the-art report, Shaker 241 p. ISBN-10: 3844037519

Self-tapping Screws as Reinforcement for Structural Timber Elements

Philipp Dietsch and Andreas Ringhofer

Abstract The use of self-tapping screws is a state-of-the-art practice in fastener and reinforcement technologies for timber structures. The high axial stiffness and load-carrying capacity of self-tapping screws, together with their easy handling, make them one of the most economical choices for applications in both fastener and reinforcement domains. This chapter focuses on mechanical models for timber members reinforced with self-tapping screws and the material properties of self-tapping screws and screwed-in threaded rods. Furthermore, the behaviour of reinforced timber members under shrinkage is discussed.

1 Introduction

Wood is a highly anisotropic material, featuring low capacities in tension and compression perpendicular to the grain as well as shear (see Fig. 1 for typical arrangements and denominations). Self-tapping screws (STS) are an efficient and economical means to compensate for these low-strength properties. They are produced by rolling or forging a wire rod around their shank, mostly over the whole length. During manufacture, the thread is hardened, leading to increased bending and torsion capacity. STS of diameters up to 14 mm and lengths up to 2000 mm, with optimized drill tips and threads, are produced today (see Fig. 2). STS as reinforcement feature threads over their full lengths (fully-threaded STS). Their application typically does not involve pre-drilling. An extension of the geometric limits of STS is given in the form of threaded rods. Threaded rods also feature screw threads over their full lengths and have diameters of up to 20 mm and lengths of up to 3000 mm. Their application

P. Dietsch (✉)
Unit of Timber Engineering, Department of Structural Engineering and Material Sciences, University of Innsbruck, Innsbruck, Austria
e-mail: philipp.dietsch@uibk.ac.at

A. Ringhofer
Institute of Timber Engineering and Wood Technology, Graz University of Technology, Graz, Austria

© RILEM 2021
J. Branco et al. (eds.), *Reinforcement of Timber Elements in Existing Structures*,
RILEM State-of-the-Art Reports 33,
https://doi.org/10.1007/978-3-030-67794-7_2

7

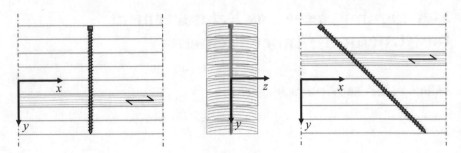

Fig. 1 Examples of arrangements of self-tapping screws as reinforcement for structural timber elements: arrangement perpendicular to the grain (left), inclined arrangement at 45° to the grain direction (right), indication of grain direction (x-direction) and perpendicular to the grain direction (y-direction)

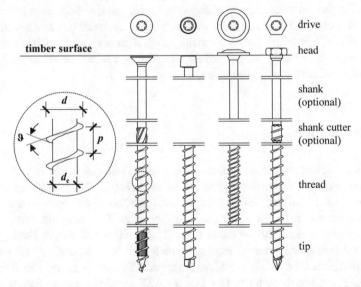

Fig. 2 Examples of different forms of screws, drill tips, threads (with and without shank cutter), and screw heads; geometric properties of screw thread (left)

involves pre-drilling with a drill diameter equal to the core diameter of the rod and a coating and/or lubricant to reduce torsional friction stresses when driving them in. Requirements for STS are given in product standards such as EN 14592 [1] in Europe. The use of other screws as specified in EN 14592 [1] is allowed, provided their applicability is proven by a technical assessment.

2 Application

Typical applications of STS as reinforcement are (see Fig. 3):

- Notched beams (Fig. 3a);
- Holes in beams (Fig. 3b);
- Glued-laminated timber (GLT) or cross-laminated timber (CLT) under high shear (Fig. 3c);
- Bearing areas/supports (Fig. 3d);
- Large dowel-type connections with slotted-in steel plates (Fig. 3e);
- Connections with a force component perpendicular to the grain (Fig. 3f);
- Carpentry joints (see separate chapter in this book);
- Curved and pitched cambered beams.

The European basis for the design of STS is the design concept given in Eurocode 5 (EN 1995-1-1 [2]) in combination with the provisions given in the European Technical Assessments (ETAs). Rules for the design, verification and detailing of STS as reinforcement are, however, not contained in the current version of EN 1995-1-1:2004 [2]. A comprehensive discussion of the state-of-the-art application and design of STS and threaded rods as reinforcement is presented in Dietsch and Brandner [3]. The Project Team SC5.T1 has prepared these design rules for the next generation of EN 1995-1-1 (see Dietsch [4]).

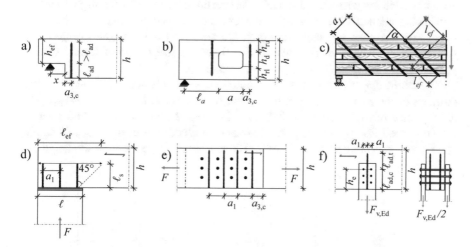

Fig. 3 Self-tapping screws used as reinforcement in notched beams (**a**), holes in beams (**b**), cross-laminated timber (**c**), supports (**d**), dowel-type connections (**e**), and connections loaded perpendicular to the grain (**f**) from [3]

3 Mechanical Models

3.1 General

Current approaches to design reinforcement against tensile stresses perpendicular to the grain assume that the stresses are entirely carried by the reinforcement [3]. This conservative assumption can be explained by the fact that the tensile strength of wood perpendicular to the grain is very low ($f_{t,90,k} = 0.5$ N/mm^2) and highly variable. In addition, moisture-induced (shrinkage) stresses reduce the tensile capacity perpendicular to the grain even more. The reinforcing effect of screws or rods is activated through the deformation of the wood around the reinforcement (here, the screw thread). Considering the low tensile strength of wood perpendicular to the grain, it becomes obvious that the corresponding deformation capacity before brittle failure (fracture) is not substantial ($\varepsilon = f_{t,90}/E_{90} = 0.5/300 = 0.17\%$).

3.2 Mechanical Models for the Unfractured State

In some reinforcement applications, it may be of interest to find out whether a proportionate distribution of stresses between the timber member and the reinforcement can be achieved in the unfractured state. This is particularly relevant if a high number of reinforcing elements is necessary to achieve the full design capacity of the timber beam, e.g. for cases of shear in which the strength (assuming shrinkage cracks $f_{v,k} \approx 2.5$ N/mm^2) is in the range of five times the magnitude of tensile strength perpendicular to the grain ($f_{t,90,k} = 0.5$ N/mm^2).

The reinforcing effect in the unfractured state is not only dependent on the relative stiffness of the reinforcement in the timber member (e.g. $E_S \cdot A_S/E_{90} \cdot A_{\text{Wood}}$) but also on the semi-rigid composite behaviour between the reinforcement and the wood material (see Fig. 4). The composite is stiffer for glued-in rods than for self-tapping screws or pre-drilled, screwed-in rods. The stiffness is dependent on the length, l_{ef},

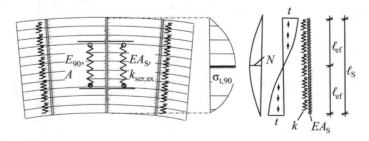

Fig. 4 Sketch of load sharing between the timber element and the reinforcement and the semi-rigid composite between the reinforcement and the wood material

the diameter, d, and the axis-to-grain angle of the screw in the wood α, (see Sect. 4 for an explicit discussion of these parameters). The distribution of stresses in the reinforcement is dependent on the abovementioned stiffness properties, as well as on the distribution of stresses in the timber member.

Mechanically, this situation can be approximated by two springs in parallel (see Fig. 4). The spring characteristics are determined on one side by the timber cross-section and the stiffness parameter associated with the stress under consideration and on the other side by the axial stiffness in combination with the semi-rigid composite behaviour (embedment modulus) of the screw in the wood.

It is possible to describe the semi-rigid composite behaviour with an embedment modulus (modulus of foundation), k, determined from the axial slip modulus, $K_{ser,ax}$, which is usually included in the technical assessments of fully threaded screws or threaded rods (see Dietsch et al. [5] for a description of the approach). Alternatively, the embedment modulus (modulus of foundation), k, can be determined from appropriate tests (see, for example, Mestek [6]).

Reinforced timber members in the unfractured state can be analysed with Theories of composite materials. In [5], an analytical approach is presented which allows calculating the efficiency of reinforcement in timber members in the unfractured state for all angles between reinforcement and wood grain. The approach is based on common theoretical concepts and constitutive equations for material properties (e.g. Lekhnitskii [7] and Klöppel and Schardt [8]). It takes account of the structural anisotropy (see Lischke [9] and Bosl [10]) of the cross-sections with reinforcement and has been extended to incorporate the semi-rigid composite behaviour between the reinforcement and the wood material. The stresses due to the forces in the timber elements are used to determine the strains in the timber elements in the direction of the reinforcement, which are in turn used to determine the proportionate distribution of stresses between the reinforcement and the timber elements (see Fig. 5). This approach has been validated by tests on reinforced glued-laminated timber (GLT) and reinforced cross-laminated timber (CLT) members (see Dietsch et al. [11]).

In some cases, the timber member is subjected to a combination of stresses, e.g. tensile stresses perpendicular to the grain and shear stresses around notches or holes, see Fig. 3a, b. Spengler [12] and Hemmer [13] have shown the negative effect of this stress interaction on the shear capacity of the wood. Screws are optimized for uniaxial loading. However, as shown by Jockwer [14], installing the screws at an angle of 45° to the grain is an efficient means to increase, for the notched timber members, not only the capacities in tension perpendicular to the grain but also in shear. Experiments reported in Danzer et al. [15] confirm this finding, showing that an arrangement of screws in the direction of the resultant of the shear stresses and the tensile stresses perpendicular to the grain (e.g. at an angle of 60° or 45°) is a good means to reinforce against these stresses around holes in beams.

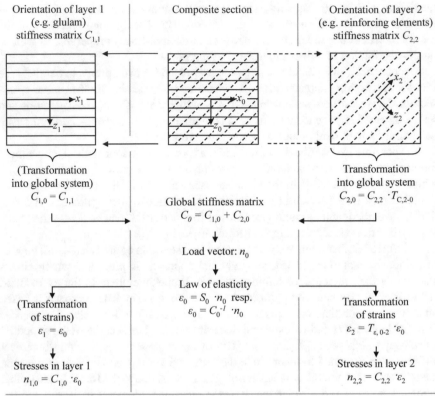

$C_{i,j}$	Stiffness matrix of layer i relating to coordinate system j
$T_{C,i-j}$	Matrix to transform the stiffness matrix from coordinate system i to j
$T_{\varepsilon,i-j}$	Matrix to transform the strains from coordinate system i to j
$n_{i,j}$	Stresses in layer i relating to coordinate system j

Fig. 5 Calculation procedure based on the structural anisotropy (in this case, for that of shear reinforcement applied at 45°, from [5])

3.3 Mechanical Models for the Fractured State

In the fractured state, the two separated parts of the timber member are held together by the reinforcement. The spring representing the wood cross-section fails and the full stresses released by the wood are concentrated on the reinforcement (see Fig. 6 and Dietsch and Kreuzinger [16]). The distribution of shear stresses (in the wood surrounding the screw thread) and axial stresses (in the reinforcement) is dependent on the stiffness of the reinforcement ($E_S \cdot A_S$) and the withdrawal stiffness of the reinforcement ($K_{ser,ax}$) in the wood (see Fig. 6). For an in-depth description of the model and mechanisms of a dowel-type fastener under pull-out loading, the interested reader is referred to the chapter "Glued-in rods as reinforcement" in this book.

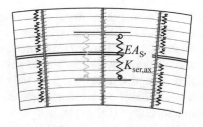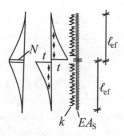

Fig. 6 Sketch of stress transfer between the timber member and the reinforcement in the fractured state (load sharing between the two timber parts disregarded for simplicity)

4 Materials

4.1 General Remarks

As introduced in Sect. 1, self-tapping screws (STS) are frequently applied, as reinforcement against stresses in the weak directions of timber members and laterally loaded dowel-type connections. Maximum efficiency is reached if the screws are arranged such that they are predominantly stressed in the axial direction. The related load-bearing performance, also denoted as withdrawal, is governed by the mechanical composite behaviour between the screw thread and its surrounding wood area. The corresponding loadbearing resistance is limited by the screw's capacity in tension or against buckling.

Typical reinforcement measures, as shown in Fig. 3, commonly consist of a comparatively small number of line-wisely or punctually arranged fasteners with threads over their full lengths (cf. Dietsch and Brandner [3]). Thus, further possible failure scenarios, according to EN 1995-1-1 [2], such as head pull-through or block shear failure (the latter of a group of screws), are barely decisive for the related design process. Consequently, in Sects. 4.2 and 4.3 we will exclusively concentrate on the aforementioned wood-screw composite behaviour (and resistance).

4.2 Parameters Influencing the Steel Properties of Screws

As mentioned in Sect. 4.1, steel failure limits the withdrawal resistance of an axially loaded STS. Consequently, in the European Technical Assessments (ETAs) for screw manufacturers, specific values for the mechanical properties of the steel, such as the characteristic tensile strength $f_{tens,k}$, the characteristic yield strength $f_{y,k}$, and the modulus of elasticity E_s, are provided for verification.

Both the characteristic tensile and yield strength values are significantly affected by the screw production process and the raw material of the steel. Nowadays, screws are either produced by special wires of stainless steel or by low-alloy carbon steel, the latter hardened and galvanised after cold-forming the geometry and only applicable for service classes 1 and 2 according to EN 1995-1-1 [2]. Based on a literature survey reported in Ringhofer [17] average tensile strengths of stainless steel screws are only about 60% of those of carbon steel screws. The relatively low tensile strengths of stainless steel screws affect their torsional strengths, thereby significantly limiting the maximum insertion length (without pre-drilling) of the screws and shortening the available thread lengths (see, for example, ETA-11/0190 [18]).

With regard to carbon steel screws, hardening not only affects the magnitude of $f_{tens,k}$, but also the screw's load-deformation behaviour and thus its yield strength and ductility. In Toblier [19], the established well-known positive relationship between the tensile strength and the core hardness (HV) was experimentally confirmed for STS. Since the hardening process varies between screw manufacturers, significantly different values for $f_{tens,k}$ are published in currently valid ETAs. For instance, $f_{tens,k}$ of carbon steel screws with an outer thread diameter, $d = 8$ mm, shows differences of 60% between minimum and maximum values published so far. In principle, screws with high $f_{tens,k}$ naturally benefit, since they have high axial loadbearing capacities (advantageous if space for the reinforcement is limited) and insertion lengths without pre-drilling. On the other hand, it can be shown that hardening greatly decreases the ductility of the fasteners, leading to more brittle behaviour if compared to conventional construction steel (cf. [19]).

Besides production and material parameters, screw steel failure is also remarkably induced by environmental conditions. High chlorine concentrations in the atmosphere (e.g. indoor swimming halls with elevated salinity or timber structures situated near a coastline and exposed to sea air), the use of wood species with high concentrations of acetic and tannic acids (e.g. Oak or Douglas fir), and service class 2 or 3 conditions, according to EN 1995-1-1 [2], accelerate electrochemical corrosion and lead to a reduction in the cross-sectional area of the screw, which might cause an exceedance of the screw's steel capacity (cf. [20]). In addition, corrosion also advances steel failure caused by hydrogen-induced stressed corrosion cracking (HSCC) (see Ringhofer [17], Landgrebe [21] and Hauptmann [22]). The abovementioned mode of failure may occur, without any prior warning, at stresses far below the tension capacity of the screw. (cf. [22]). Currently, barely little is known regarding HSCC of STS. Nevertheless, material conditions that favour this phenomenon, such as the use of carbon steel screws with high tensile capacity ($f_{tens,k} > 1000$ MPa), in combination with the aforementioned environmental conditions should be considered in the design process.

Even though specific stainless steel grades are considered appropriate for this purpose (e.g. steel grade K5 according to FprEN 14592 [23]), related application should be planned carefully after consulting the specific screw manufacturer.

None of the aforementioned parameters has a remarkable influence on the screw's modulus of elasticity, E_s.

4.3 Parameters that Determine Wood-Screw Composite Behaviour

4.3.1 Overview

With respect to the withdrawal behaviour of an axially loaded STS, the corresponding failure mechanism can be reduced to local failure in the surrounding wood (shear in the different planes in combination with tension/compression perpendicular to the grain, cf. Ringhofer [17] and Hübner [24]) interacting as a composite with the inserted threaded part of the screw. Strength and stiffness withdrawal properties, essential for the modelling and verification of STS applied as reinforcement (cf. Bejtka [25], Mestek [6], Dietsch [26]), depend on various geometrical, physical and mechanical parameters. One way to classify these parameters is shown in Fig. 7.

The current state-of-the-art concerning the effect of these parameters on withdrawal properties is summarised in the following subsections. Both strength and stiffness are denoted as $X = \{f_{ax}, k_{ser,ax}\}$ and determined according to Eq. (1):

$$X = \frac{Y}{d \cdot \pi \cdot l_{ef}}, \tag{1}$$

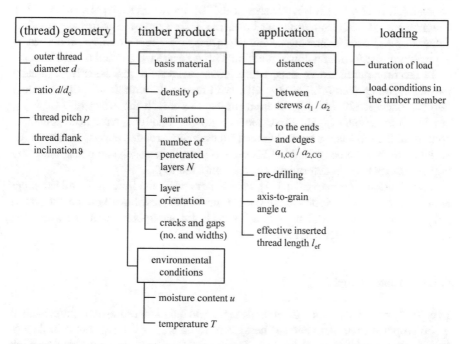

Fig. 7 Parameters that determine the withdrawal properties of STS, according to Ringhofer et al. [27]

where $Y = \{F_{ax}, K_{ser,ax}\}$ is the set of withdrawal capacity (in N) and stiffness (in N/mm) respectively and d and l_{ef} are the screw's outer thread diameter and effective inserted thread length excluding the tip, both in mm.

4.3.2 Screw (Thread) Geometry

As given in Fig. 7, the thread geometry of modern STS can be described by the outer thread diameter, d, the ratio of d to the inner thread diameter d_c (sometimes also denominated as d_1 or the root diameter), the thread pitch, p, and the flank inclination, ϑ. Current standards and regulations covering the design and application (CE-labelling) of the STS limit d and d/d_c to certain ranges: 2.4 mm $\leq d \leq$ 24 mm (according to EN 14592 [1]) and $0.6\,d \leq d_c \leq 0.75\,d$ (according to EN 1995-1-1 [2]), $0.6\,d \leq d_c \leq 0.9\,d$ (according to EN 14592 [1]), and $0.5\,d \leq d_c \leq 0.9\,d$ (according to EAD 130118-00-0603 [28]). In case of p and ϑ no related restriction was found. Typical values for the thread pitch and the flank inclination are $p \approx 0.5\,d$ and $\vartheta \approx$ 40° (cf. [17]).

Like for dowel-type fasteners, all empirical regression models published so far to determine the withdrawal strength of the modern STS take account of the outer diameter d, which represents the effect of the thread characteristics on the withdrawal strength (cf. Blaß et al. [29], Pirnbacher et al. [30], Frese and Blaß [31], Hübner [24], Ringhofer et al. [32] and Brandner et al. [33]). The reported regressive trend of decreasing f_{ax} with increasing d, indicating the size-effect of wood properties, is expressed in the form of a power function with a negative exponent in most cases.

In recently published models, nearly equal values for this negative exponent (≈ -0.34) are applied (cf. [24, 29, 32]). With respect to withdrawal stiffness, only two approaches, [29 and 32], wherein for each approach the effect of d on $k_{ser,ax}$ is taken into account (again expressed as a power function), have been found. In contrast to the models for the withdrawal strength, both related exponents show high variation in their values (-0.8 in [29] vs. -1.7 in [32]), which can be explained by higher uncertainties in determining $k_{ser,ax}$ compared to f_{ax}.

Investigations summarised in [31], Pirnbacher and Schickhofer [34] and Brandner et al. [33] indicate no significant impact of the other screw thread parameters, d/d_c, p and ϑ, on the withdrawal strength of the STS, for parameter variations within the aforementioned geometrical restrictions.

4.3.3 Wood Material

In general, the wood density, ρ, serves as a material indicator representing mechanical wood properties (strength and stiffness) used to determine the resistance of dowel-type fasteners (cf. EN 1995-1-1 [2]). The reason for this is the comparatively high correlation between ρ and the fastener's embedment or withdrawal strength.

For the STS, the positive and pronounced relationship between ρ and the withdrawal properties is commonly described by a power function (see [23, 29, 31, 32] for

f_{ax} and [29, 32] for $k_{ser,ax}$). For the withdrawal strength, the exponent k_ρ considered in the aforementioned approaches shows a comparatively high variation between 0.75 and 1.60. This can be explained by the fact that the parameters, d and the axis-to-grain angle α, significantly affect the relationship between f_{ax} and ρ. One approach with k_ρ as a function of d and α (instead of a constant value for k_ρ as used in previous works) is provided in [32], increasing the model's predictability to a certain extent. In the case of the withdrawal stiffness, the approaches in [29 and 32] also consider k_ρ as a constant factor (0.2 in [29] and 0.75 in [32]); a functional relationship of k_ρ with the other parameters has not been described so far.

Research (Reichelt [35]) on the STS applied in laminated timber products, such as glued-laminated timber (GLT) or cross-laminated timber (CLT), considers the number of layers N penetrated by the screw as having a significant effect on its withdrawal properties. As demonstrated in [35], this effect, which is in the form of increasing strength and stiffness with increasing N, not only concerns the related characteristic values (5%) but also the mean values. A corresponding stochastic approach, in the form of $f_{ax,N} = k_{sys}(N) \cdot f_{ax,N=1}$, is derived in Ringhofer et al. [36], whose recommended k_{sys} values for the mean and characteristic properties are given in Table 1.

In contrast to the number of penetrated layers, the specific orientation (unidirectional for GLT; orthogonal for CLT), which is another essential parameter for timber product classification, has no significant effect on the withdrawal properties (cf. [35, 36]). Since the possibility of screw insertion in gaps or cracks cannot be excluded when applying STS for the reinforcement of CLT structures, a corresponding loss in strength and stiffness has to be considered. Based on the findings by Brandner et al. [37], Silva et al. [38] and Ringhofer [17], the decreasing withdrawal properties $X_{CLT,gap}$ of a screw inserted in gaps with varying widths (w_{gap}), where the number of gaps is given by n_{gap}, can be described according to Eq. (2):

$$X_{CLT,gap} = \frac{n_{gap}}{N}\left(k_{gap} - 1 + \frac{N}{n_{gap}}\right) \cdot X_{CLT}, \tag{2}$$

where X_{CLT} is the base value of the screw's withdrawal property in CLT, $k_{gap} = P_{red}/P_{tot}$, P_{red} and P_{tot} are its reduced (effective) and total circumference of the outer thread ($\pi \cdot d$), and n_{gap} is the number of gaps the screw is inserted into.

For practical application, the approaches published in [17, 32] and in Uibel and Blaß [39] take into account the existence and the negative effect of gaps in CLT panels on the withdrawal performance of screws randomly inserted in the CLT.

Table 1 $k_{sys,mean}$ and $k_{sys,k}$ values proposed for GLT and CLT as functions of N (number of layers penetrated), according to [36]

N	1	2	3	4	5	6	7	8	9	10
$k_{sys,mean}$	1.00	1.05	1.07	1.09	1.10	1.11	1.12	1.12	1.13	1.13
$k_{sys,k}$	1.00	1.06	1.10	1.12	1.13	1.14	1.15	1.16	1.17	1.17

4.3.4 Environmental Conditions

As mentioned before, the withdrawal behaviour of the STS is governed by the wood area surrounding the thread. The varying moisture content u of the wood, caused by a change in environmental conditions (temperature and relative humidity), considerably affects the withdrawal strength and stiffness of the screw in the wood. According to [32] and Ringhofer et al. [40], which both determined the withdrawal properties of STS in solid timber (ST), GLT and CLT from Norway spruce with a high range of u, the corresponding behaviour is as follows:

$$\frac{X_u}{X_{ref}} = \begin{cases} 1.00 & 8\% \le u \le 12\% \\ 1.00 - k_{mc} \cdot (u - 12) & u \ge 12\% \end{cases}, \tag{3}$$

where $\{X_u, X_{ref}\}$ are the withdrawal properties at $u = \{u, 12\%\}$ and $k_{mc} = \{0.034, 0.017\}$ for $\{f_{ax}, k_{ser,ax}\}$ are the gradients describing the linear decrease in X with increasing u for values exceeding 12%. In [24], a similar behaviour for the withdrawal strength in hard timber products was found ($k_{mc} = 0.024 - 0.027$).

It is worth mentioning that in both [24, 40], the target moisture content of the wood specimens was reached by *static* climate conditioning. Studies on the impact of *cyclic* climate variation on screw withdrawal properties are comparatively scarce. One such scare study is the work by Silva et al. [41], in which they conducted cyclic changes in the moisture content (4 cycles from 30% r.h. to 90% r.h. at $T = 20\,°C$, 324 days of storage, $u_{start} = u_{finish} \approx 13 - 14\%$) of GLT and CLT specimens assembled with self-tapping screws of $d = 8$ mm and tested after conditioning. Based on their observations, and presupposing that $u_{start} = u_{finish}$, a clear tendency regarding the impact of variations in the cyclic moisture content variation on withdrawal properties could not be found so far. With regard to the loadbearing performance of STS applied in timber members with varying temperature T, results reported in [30] indicate no remarkable effect of T (between $-20\,°C$ and $50\,°C$) on withdrawal strength f_{ax}. In order to avoid wood splitting, it is recommended to install the screw at $T \ge 5\,°C$.

4.3.5 Application

The minimum spacing and edge and end distances a_i of dowel-type fasteners are very important for the design of connections and reinforcement. Accurate spacing and distances, on the one hand, help to avoid wood splitting during installation and, on the other hand, to prevent brittle failure of the timber member surrounding a group of screws (block or plug shear). Table 2 compares the regulations, according to EN 1995-1-1 [2] and a selected ETA [18], on the size of a_i for (predominantly) axially loaded STS. These regulations represent the current state of knowledge on these parameters.

Determination of a_i for ETAs was recently carried out in the form of a screwing-in test, which included the optical assessment of the corresponding crack formation

Table 2 Minimum distances of axially loaded self-tapping screws according to EN 1995-1-1 [2] and ETA-11/0190 [18]

Type	Between screws in the direction of the grain	Between screws perpendicular to the direction of the grain	To the end grain	To the edge
	a_1	a_2	$a_{1,CG}$	$a_{2,CG}$
EN 1995-1-1 [2]	$7d$	$5d$	$10d$	$4d$
ETA-11/0190 [18]	$5d$	$5d^a$	$5d$	$4d$

^{a}Can be reduced to $2.5d$ if $a_1 \cdot a_2 \geq 25d^2$

(cf. EAD 130118-00-0603 [28]). Since this procedure was developed for the assessment of screw application in an unloaded state, the potential effect of minimum spacing on the withdrawal performance of a screwed connection or reinforcement could not be taken into account. Plieschounig [42] conducted withdrawal tests with two simultaneously loaded screws inserted in a solid wood specimen with varying spacing ($a_1 = 2d - 14d$, $a_2 = 2d - 5d$) and $\alpha = 90°$. Related findings showed no reduction in withdrawal strength when $\{a_1, a_2\} \geq \{7d, 3d\}$, indicating that values published in currently valid ETAs overestimate the bearing capacity for this failure mode. Plüss and Brandner [43] confirmed the aforementioned value for $a_{1,min}$ at $\alpha = 90°$ and showed that the effect of a_1 on withdrawal strength decreases with decreasing α: $\{a_{1,\alpha=45°}; a_{1,\alpha=0°}\} \geq \{5d; 2.5d\}$.

With regard to the minimum distances to the ends and edges of the timber member and according to Gatternig [44], the withdrawal strength is not affected if $\{a_{1,CG};$ $a_{2,CG}\} \geq \{3d, 1d\}$ and $\alpha = 90°$, indicating that values given in Table 2 are sufficient in this case. It is worth mentioning that in none of [42–44], the screw heads were sunk into the timber member, which would have increased crack formation and might have required higher spacing, thereby affecting $a_{1,CG}$. So far, there are no studies focused explicitly on the effect of varying minimum spacing on withdrawal stiffness.

STS application requires pre-drilling to avoid screw failure due to torsion, especially in cases of timber products made of species of densities ρ exceeding 550 kg/m^3. According to [17, 30, 35], Ringhofer and Schickhofer [45] and Brandner et al. [33], this requirement has no remarkable effect on withdrawal properties, presupposing that the borehole diameter d_{PD} does not exceed 80% of d. Appropriate values for d_{PD}, which vary in dependence of d and the wood species are commonly given in related ETAs (see, for example [18]).

Due to the natural anisotropy of wood, its mechanical properties considerably depend on the angle between the stress and grain directions. Since this dependence also concerns the withdrawal behaviour of axially loaded STS, α as the axis-(load)-to-grain angle is one of the most important parameters that affect the design process of STS. Like in other dowel-type fasteners, the effect of α on the withdrawal strength of STS is commonly described by k_{90}, which is the ratio of f_{ax} to $\alpha = 90°$ and is part of a functional relationship covering the bandwidth between both limits of α. In each of [2, 24, 29–33, 37, 39], the corresponding description of withdrawal strength (decreasing f_{ax} with decreasing α) is provided in different ways, including a model similar to that of Hankinson [46] (see [2, 29, 30, 39]), a bilinear approach (no change for (30)$45° < \alpha \leq 90°$, followed by a linear decrease for $\alpha \leq 45°$, as seen in [25, 32 and 33] and several ETAs), and also a polynomial function as provided in [37]. Unlike k_ρ, all the aforementioned models consider k_{90} a constant value, with values for the models ranging between 1.20 and 1.56. For withdrawal stiffness, in [32] an inverse relationship (increasing $k_{ser,ax}$ with decreasing α) is observed and consequently described by a bilinear approach, with $k_{90} = 0.75$.

With regard to l_{ef}, herein defined as effective inserted threaded part of the screw excluding its tip, only the model given in [29] (as the basis of the design equation in EN 1995-1-1 [2]) considers a related influence on withdrawal strength. In the study, a slight decrease in f_{ax} with increasing l_{ef} is modelled by a power function with a negative exponent -0.10. In the works [25, 29–33], no related term is applied, indicating a linear relationship of l_{ef} to the withdrawal capacity F_{ax}, which is seen as the current state of knowledge on this parameter. Withdrawal stiffness, however, is significantly affected by varying insertion lengths. In each of [29, 32], a remarkable decrease in $k_{ser,ax}$ with increasing l_{ef} is observed and described by a power function, with an identical negative exponent -0.60. For axially loaded threaded rods, a similar relationship between $k_{ser,ax}$ and l_{ef} was found by Stamatopoulos and Malo [47], while Blaß and Krüger [48] experimentally determined a different behaviour, i.e. increasing $k_{ser,ax}$ with increasing l_{ef}. One possible explanation of these deviating findings is seen in the fact that, currently, there is no clear definition as regards the determination of withdrawal stiffness.

4.3.6 Loading

According to Colling [49, 50] and [25], there are no effects on withdrawal properties if the axial load transferred by the screw changes between tension and compression.

Equal to the mechanical properties of the wood in general, the withdrawal performance of STS significantly depends on the load duration (DoL) and creep effects. Because great efforts are needed to determine the corresponding factors k_{mod} and k_{def}, according to EN 1995-1-1 [2], related investigations focusing on axially loaded STS are scarce. An experimental study carried out by Pirnbacher and Schickhofer [51] found that screws at $\alpha \geq 45°$ showed long-time behaviour comparable to that of the timber member they were placed in, while parallel-to-grain insertion led to an immediate loss of bearing capacity only a few minutes after the loads were applied. Consequently, in ETAs which consider this form of screw application, withdrawal strength at $\alpha = 0°$ is limited to 30% of that at $\alpha = 90°$ ($k_{90} = 3.33$). Surprisingly, latest research activities on STS inserted in solid timber made of soft- and hardwood in parallel to grain ($\alpha = 0°$) confirm the k_{mod}-factors, which are recommended in [2], also for STS, cf. Brandner et al. [52]. This led to an ongoing discussion regarding the future regulation of this important parameter for the design and application of STS.

With respect to the axial stiffness, experiments reported in Dietsch [26] indicate that the first loading cycles lead to an increase in stiffness of the semi-rigid composite between screw thread and wood.

As summarised in [3], reinforcements are commonly located in wood areas stressed in bending or shear. In order to verify if corresponding stresses affect the axial loadbearing capacity of STS, Kolany [53] carried out experiments that included withdrawal tests in the compression, tension and shear zones of timber beams loaded in bending. Based on his findings, differences in withdrawal strength due to screw location and loading level are considered negligible.

5 Effect of Reinforcement on Shrinkage Stresses in Timber

5.1 General

Changes in moisture content lead to strains of different magnitude across the timber cross-section due to the associated shrinkage or swelling of the wood. In the case of free deformation, there is an equilibrium of the tensile and compressive stresses perpendicular to the grain over the cross-section. If the free deformation of the cross-section is prevented by restraining forces, e.g. the semi-rigid composite behaviour between the wood material and the thread of the reinforcement, the equilibrium of the stresses perpendicular to the grain is impeded. The magnitude of moisture-induced stresses depends on the difference between the strains of the timber cross-section and the restraining elements (see Fig. 8). This difference in strains results in

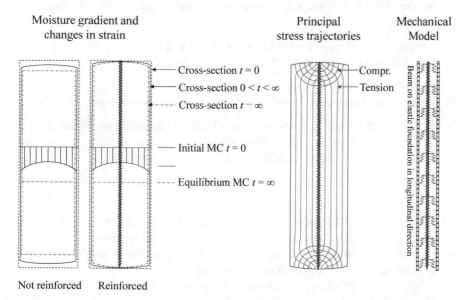

Fig. 8 Deformed shape of a timber cross-section (not reinforced and reinforced), principal stress trajectories, and mechanical model of a reinforced timber cross-section under shrinkage

stresses of higher magnitude and eventually in a few but deep shrinkage cracks. A crack distribution, known from reinforced concrete structures under shrinkage, does not occur. Dietsch [26, 54] describe three factors to explain this fact: (1) the high variability of strength perpendicular to the grain within a reinforced timber member (first crack appearing at the "weakest link"); (2) the small ratio of strength to stiffness of the material necessary to achieve a quick stress transfer at little deformation in the direct vicinity of the crack; (3) the large material cover, i.e. the distance between the surface of the timber member and the reinforcement.

5.2 Experimental and Analytical Investigations

Research on reinforced timber members under changing moisture content u is scarce [54–58]. In [56–58], long-term experiments on reinforced timber members placed in climate chambers were carried out. The challenge in such experiments is how to apply climate cycles that do not lead to high moisture gradients, thus making it easier to differentiate between crack formation due to the moisture gradient or due to the reinforcement restraining free shrinkage. In [54] (see also [26]), pull-pull tests on the steel rods of reinforced glued-laminated timber specimens were carried out. These are justified by the finding that, although the nature of the strain (shrinkage strain or strain due to an externally applied tensile load) is different from one another, the stress distribution in the timber member, which results from the interaction between the wood and the reinforcement, is comparable. Thus results in [54, 58] emphasize that:

- the restraining effect of the reinforcement leads to a few but deep cracks;
- the restraining effect increases with:

  - increasing height of the reinforced section;
  - decreasing distance between the reinforcement (screws or rods).

- in the case of a single reinforcing element placed perpendicular to the grain of one timber member ($h > 1$ m), a reduction of $u = 3$–4% around one single screw or rod can lead to critical stress levels with respect to cracking.

Analytical considerations and finite-element simulations in [54] (see also [26]) indicate that:

- a change of angle between the reinforcement and wood grain from 90° to 45° reduces the shrinkage stresses perpendicular to the grain by about half (Fig. 9);
- a reduction in the distance between the reinforcing elements results in a lower tolerable reduction in wood moisture content;
- the reinforcement elements should be placed at the centre of the width of the cross-section.

Recently, above given indications were confirmed and specified by Danzer et al. [59, 60]. Via extensive experimental and numerical investigations it was shown

$a_1 = 750$ mm

Fig. 9 Distribution of stresses perpendicular to the grain (vertical stresses normalized) in a GLT member ($h/b = 1.0/0.2$ m) reinforced with steel rods ($d = 16$ mm) at angles of 90° and 45° [54]

that the onset of cracking is influenced in particular by the quantity of reinforcing elements, their inclination to grain direction, the member height and the axial withdrawal stiffness of the reinforcement. All four influencing factors have a negative effect on the formation of cracks with increasing magnitude. Crack formation can be expected at differential wood moisture contents $\Delta u \approx 3\% - 5.5\%$, depending on the type, quantity and arrangement of the reinforcement. The load-carrying capacities of specimen with reinforced holes or notches after shrinkage ($\Delta u = 2.5\% - 5.3\%$) reduced to 65%–83%, compared to tests on reinforced specimen before shrinkage. In both cases, internal reinforcement at 45° showed a higher reinforcing effect combined with a lower restraining effect.

6 Conclusions

The use of self-tapping screws (STS) is state-of-the-art practice in fastener and reinforcement technology for structural timber elements. Their high axial stiffness and load-carrying capacity, as well as their easy handling, make them one of the most economical choices for applications in both reinforcement and fastener domains. Their axial strength parameters are well described and investigated. Apart from the wood density ρ, the screw outer thread diameter d, and the axis-to-grain angle α, all of which were taken into account in all known prediction models for f_{ax}, the timber member's moisture content u, the number of penetrated layers N, and the spacing a_i between screws also have major impacts on the magnitude of f_{ax}. There is the need for further investigation of the load duration effects of axially loaded self-tapping screws, with a special focus on parallel-to-grain insertion. The efficiency of screws as reinforcement in unfractured timber members is highly affected by the stiffness of the screw thread and wood composite. In this domain, results are still heterogeneous. The key to more comparable results would be a clear definition of how to determine the composite stiffness (withdrawal stiffness) from tests.

The reinforcing effect of self-tapping screws is reliable in cases where they are installed in a timber member with a moisture content close to its equilibrium moisture content in use. Precautions should be taken against potential shrinkage of the timber member ($\Delta u \geq 3\%$). These include applying the screws at larger distances apart and/or at lower angles to the grain as well as reducing the height of the reinforced timber section. In addition, the screws should be designed to carry the full stresses under consideration. In the case of reinforcement for tensile stresses perpendicular to the grain in areas with considerable shear stresses, the shear stresses should also be directed to the reinforcement. A common alternative is to substitute STS with wood-based panels that are glued to the sides of the beam, i.e. that also exhibit a damping effect in the case of changing wood moisture. Recent studies show the potential of an arrangement of screws in the direction of the resultant of a stress combination, also with respect to the reduced restraining effect during shrinkage of the reinforced timber member.

References

1. EN 14592:2012 (2012) Timber structures—dowel type fasteners—requirements. CEN, Brussels
2. EN 1995-1-1:2004-11 (2004) Eurocode 5: design of timber structures—Part 1-1: general—common rules and rules for buildings. CEN, Brussels
3. Dietsch P, Brandner R (2015) Self-tapping screws and threaded rods as reinforcement for structural timber elements—a state-of-the-art report. Constr Build Mater 97:78–89. https://doi.org/10.1016/j.conbuildmat.2015.04.028
4. Dietsch P (2019) Eurocode 5:2022 - A new section on reinforcement of timber structures. In: SHATiS international conference on structural health assessment of timber structures, Guimarães, Portugal
5. Dietsch P, Kreuzinger H, Winter S (2013), Design of shear reinforcement for timber beams. CIB-W18/46-7-9, Meeting 46 of the working commission W18-timber structures. CIB, Vancouver, Canada
6. Mestek P (2011) Punktgestützte Flächentragwerke aus Brettsperrholz (BSP)—Schubbemessung unter Berücksichtigung von Schubverstärkungen. Dissertation, Technische Universität München
7. Lekhnitskii SG (1968) Anisotropic plates (translated from Russian by Tsai SW and Cheron TG). Gordon and Breach Science Publishers, New York, London, Paris
8. Klöppel K, Schardt R (1960) Systematische Ableitung der Differentialgleichungen für ebene anisotrope Flächentragwerke. Stahlbau 29(2):33–43
9. Lischke N (1985) Zur Anisotropie von Verbundwerkstoffen am Beispiel von Brettlagenholz, Fortschritts-Bericht VDI, Reihe 5, Nr. 98, VDI-Verlag, Düsseldorf
10. Bosl R (2002) Zum Nachweis des Trag- und Verformungsverhaltens von Wandscheiben aus Brettlagenholz. Dissertation, Universität der Bundeswehr, München
11. Dietsch P, Mestek P, Winter S (2012) Analytischer Ansatz zur Erfassung von Tragfähigkeitssteigerungen infolge von Schubverstärkungen in Bauteilen aus Brettschichtholz und Brettsperrholz. Bautechnik 89(6):402–414
12. Spengler R (1982) Festigkeitsverhalten von Brettschichtholz unter zweiachsiger Beanspruchung, Teil 1—Ermittlung des Festigkeitsverhaltens von Brettelementen aus Fichte durch Versuche, Berichte zur Zuverlässigkeitstheorie der Bauwerke, Heft 62, LKI der TU München
13. Hemmer K (1984) Versagensarten des Holzes der Weißtanne (Abies Alba) unter mehrachsiger Beanspruchung. Dissertation, TH Karlsruhe
14. Jockwer R (2014) Structural behaviour of glued laminated timber beams with unreinforced and reinforced notches. Dissertation, ETH Zurich
15. Danzer M, Dietsch P, Winter S (2017) Round holes in glulam beams arranged eccentrically or in groups. INTER/50-12-6, International Network on Timber Engineering Research INTER, Meeting 4, Kyoto, Japan

16. Dietsch P, Kreuzinger H (2020) Dynamic effects in reinforced beams at brittle failure—evaluated for timber members. Eng Struct 209
17. Ringhofer A (2017) Axially loaded self-tapping screws in solid timber and laminated timber products. Dissertation, Graz University of Technology
18. ETA-11/0190 (2018) Adolf Würth GmbH & Co. KG: self-tapping screws for use in timber constructions. European Technical Assessment, DIBt
19. Toblier L (2014) Untersuchung der wasserstoffinduzierten Versprödung von verzinkten, hochfesten Holzschrauben. Bachelor Thesis, Graz University of Technology
20. Informationsdienst Holz (2013) Spezial – Korrosion metallischer Verbindungsmittel in Holz und Holzwerkstoffen. Fraunhofer Institute for Wood Research, Braunschweig
21. Landgrebe R (1993) Wasserstoffinduzierte Sprödbruchbildung bei hochfesten Schrauben aus Vergütungsstählen. Dissertation, TU Darmstadt
22. Hauptmann R (2016) Experimentelle Untersuchung der wasserstoffinduzierten Versprödung von verzinkten, hochfesten Holzschrauben unter extremen Umgebungsbedingungen. Bachelor Thesis, Graz University of Technology
23. FprEN 14592:2019 (2019) Timber structures—dowel type fasteners—Requirements. CEN, Brussels
24. Hübner U (2013) Withdrawal strength of self-tapping screws in hardwoods. CIB-W18/46-7-4, Meeting 46 of the Working Commission W18-Timber Structures, CIB, Vancouver, Canada
25. Bejtka I (2005) Verstärkung von Bauteilen aus Holz mit Vollgewindeschrauben. Dissertation, Band 2 der Reihe Karlsruher Berichte zum Ingenieurholzbau, KIT Scientific Publishing, Karlsruhe
26. Dietsch P (2012) Einsatz und Berechnung von Schubverstärkungen für Brettschichtholzbauteile. Dissertation, Technische Universität München
27. Ringhofer A, Grabner M, Brandner R, Schickhofer G (2014) Die Ausziehfestigkeit selbstbohrender Holzschrauben in geschichteten Holzprodukten. Doktorandenkolloquium Holzbau „Forschung und Praxis, Stuttgart (Germany)
28. EAD 130118-00-0603 (2019) Screws and threaded rods for use in timber structures. EOTA
29. Blaß HJ, Bejtka I, Uibel T (2006) Tragfähigkeit von Verbindungen mit selbstbohrenden Holzschrauben mit Vollgewinde. Band 4 der Reihe Karlsruher Berichte zum Ingenieurholzbau, KIT Scientific Publishing, Karlsruhe
30. Pirnbacher G, Brandner R, Schickhofer G (2009) Base parameters of self-tapping screws. CIB-W18/42-7-1, Meeting 42 of the Working Commission W18-Timber Structures, CIB, Duebendorf, Switzerland
31. Frese M, Blaß HJ (2009) Models for the calculation of the withdrawal capacity of self-tapping screws. CIB-W18/42-7-3, Meeting 42 of the Working Commission W18-Timber Structures, CIB, Duebendorf, Switzerland
32. Ringhofer A, Brandner R, Schickhofer G (2015) A universal approach for withdrawal properties of self-tapping screws in solid timber and laminated timber products. INTER/48-7-1, International Network on Timber Engineering Research INTER, Meeting 2, Sibenik, Croatia
33. Brandner R, Ringhofer A, Reichinger T (2019) Performance of axially-loaded self-tapping screws in hardwood: properties and design. Eng Struct 188:677–699
34. Pirnbacher G, Schickhofer G (2007) Schrauben im Vergleich – eine empirische Betrachtung. 6. Grazer Holzbau-Fachtagung, Graz (Austria)
35. Reichelt B (2012) Einfluss der Sperrwirkung auf den Ausziehwiderstand selbstbohrender Holzschrauben. Master Thesis, Graz University of Technology
36. Ringhofer A, Brandner R, Schickhofer G (2015) Withdrawal resistance of self-tapping screws in unidirectional and orthogonal layered timber products. Mater Struct 48(5):1435–1447
37. Brander R, Ringhofer A, Grabner M (2018) Probabilistic models for the withdrawal behavior of single self-tapping screws in the narrow face of cross laminated timber (CLT). Eur J Wood Wood Prod 76:13–30
38. Silva C, Ringhofer A, Branco JM, Lourenco PB, Schickhofer G (2014) Influence of moisture content and gaps on the withdrawal resistance of self-tapping screws in CLT. 9°Congresso Nacional de Mecânica Experimental, Aveiro (Portugal)

39. Uibel T, Blaß HJ (2007) Edge joints with dowel type fasteners in cross laminated timber. CIB-W18/40-7-2, Meeting 40 of the Working Commission W18-Timber Structures, CIB, Bled, Slovenia
40. Ringhofer A, Grabner M, Silva C, Branco JM, Schickhofer G (2014) The influence of moisture content variation on the withdrawal capacity of self-tapping screws. Holztechnologie 55(3):33–40
41. Silva C, Branco JM, Ringhofer A, Lourenco PB, Schickhofer G (2016) The influences of moisture content variation, number and width of gaps on the withdrawal resistance of self-tapping screws inserted in cross laminated timber. Constr Build Mater 125:1205–1215
42. Plieschounig S (2010) Ausziehverhalten axial beanspruchter Schraubengruppen. Master thesis, Graz University of Technology
43. Plüss Y, Brandner R (2014) Untersuchungen zum Tragverhalten von axial beanspruchten Schraubengruppen in der Schmalseite von Brettsperrholz (BSP). 20. Internationales Holzbau-Forum (IHF), Garmisch-Partenkirchen (Germany)
44. Gatternig W (2010) Untersuchung der Randabstände bei selbstbohrenden Holzschrauben. Research report, Graz University of Technology
45. Ringhofer A, Schickhofer G (2014) Influencing parameters of the experimental determination of the withdrawal capacity of self-tapping screws. In: 13th world conference on timber engineering (WCTE), Quebec (Canada)
46. Hankinson RL, Investigation of crushing strength of spruce at varying angles of grain. U.S. Air service information circular, vol 3, no 259. Washington D.C., US Air Service, Materials Section Paper No. 130
47. Stamatopoulos H, Malo KA (2015) Characteristic withdrawal capacity and stiffness of threaded rods. INTER/48-7-2, International Network on Timber Engineering Research INTER, Meeting 2, Sibenik, Croatia
48. Blaß HJ, Krüger O (2010) Schubverstärkung von Holz mit Holzschrauben und Gewindestangen. Band 15 der Reihe Karlsruher Berichte zum Ingenieurholzbau, KIT Scientific Publishing, Karlsruhe
49. Colling F (2001) Erhöhung der Querdruckfestigkeit von Holz – Forschungsergebnisse FH Augsburg, Teil 1. mikado 10:66–69
50. Colling F (2001) Erhöhung der Querdruckfestigkeit von Holz – Forschungsergebnisse FH Augsburg, Teil 2. mikado 11:62–66
51. Pirnbacher G, Schickhofer G (2012) Zeitabhängige Entwicklung der Traglast und des Kriechverhaltens von axial beanspruchten, selbstbohrenden Holzschrauben. Research Report, holz.bau forschungs gmbh, Graz University of Technology
52. Brandner R, Ringhofer A, Sieder R (2019) Duration of load effect on axially-loaded self-tapping screws inserted parallel to grain in soft- and hardwood. INTER/52-9-1, International Network on Timber Engineering Research INTER, Meeting 6, Tacoma, USA
53. Kolany GHE (2009) Ausziehwiderstand von selbstbohrenden Holzschrauben in Bauteilen unter Biegebeanspruchung. Master's thesis, Graz University of Technology
54. Dietsch P (2017) Effect of reinforcement on shrinkage stresses in timber members. Constr Build Mater 150:903–915
55. Hollinsky K (1992) In Brettschichtholz eingeklebte Stabelemente: Verhalten bei wechselnden klimatischen Bedingungen, Holzforschung und Holzverwertung Nr. 1
56. Ehlbeck J, Belchior-Gaspard P, Gerold M (1992) Eingeleimte Gewindestangen unter Axial-belastung bei Übertragung von großen Kräften und bei Aufnahme von Querzugkräften in Biegeträgern – Teil 2: Einfluß von Klimaeinwirkung und Langzeitbelastung. Forschungsbericht der Versuchsanstalt für Stahl, Holz und Steine, Universität Karlsruhe
57. Blaß HJ, Krüger O (2010) Schubverstärkung von Holz mit Holzschrauben und Gewindestangen. Band 15 der Reihe Karlsruher Berichte zum Ingenieurholzbau, vol 15. KIT Scientific Publishing, Karlsruhe
58. Wallner B (2012) Versuchstechnische Evaluierung feuchteinduzierter Kräfte in Brettschichtholz verursacht durch das Einbringen von Schraubstangen. Master's thesis. Institute of Timber Engineering and Wood Technology, Graz University of Technology

59. Danzer M, Dietsch P, Winter S (2020) Shrinkage behaviour of reinforced Glulam members. INTER/53-12-1, International Network on Timber Engineering Research INTER, Meeting 7, Online
60. Danzer M, Dietsch P, Winter S (2019) Verhalten verstärkter Brettschichtholzbauteile unter Schwindbeanspruchung. Forschungsbericht. Lehrstuhl für Holzbau & Baukonstruktion, TU München

Glued-in Rods as Reinforcement for Timber Structural Elements

Robert Jockwer and Erik Serrano

Abstract Glued-in rods are important connecting and reinforcing elements in modern timber engineering used in new and existing timber structures. The complex stress distribution along the bondline between rod and wood depends on the type and properties of the adhesive and the type of load application. In this chapter, different models for the determination of the shear stress distribution along the bondline are discussed and their effects on the stress distribution and strength of the glued-in rod are evaluated. Important points on how to enhance the load-carrying capacity and reach best structural capacity, as well as the ductile failure behaviour, are discussed.

1 Introduction

Glued-in rods (GiRs) are widely used in modern timber structures for connection and reinforcement purposes. Their main benefits are their high load-carrying capacity and high stiffness. The development of GiR started from the need to transfer high loads into elements for modern timber structures, with the first experimental studies conducted in the 1970s [1–3]. Recent developments have aimed at avoiding stress peaks at the surface of the timber elements by shifting the adhered zone into the element to achieve more ductile failure [4, 5].

The compound of wood, adhesive and rod requires a good quality and manufacturing control to prevent brittle failure behaviour provoked by stress peaks occurring at the bondline. When glued-in rods are produced in a factory, a high precision can normally be ensured. Highly loaded connections and reinforced sections with

R. Jockwer (✉)
Division of Structural Engineering, Department of Architecture and Civil Engineering (ACE), Chalmers University of Technology, Göteborg, Sweden
e-mail: robert.jockwer@chalmers.se

E. Serrano
Division of Structural Mechanics, Lund University, Lund, Sweden

29

J. Branco et al. (eds.), *Reinforcement of Timber Elements in Existing Structures*,
RILEM State-of-the-Art Reports 33,
https://doi.org/10.1007/978-3-030-67794-7_3

multiple GiRs require special attention to failure in the surrounding timber cross-section and block shear failure of groups of GiRs. In addition, the risk of progressive failure in pull-out for each group of GiRs should also be assessed [6–10].

The complexity of the structural behaviour of wood, adhesive and rod was subject of numerous studies and led to a variety of different design proposals. A comprehensive overview of the geometry and various parameters that affect the load-carrying behaviour of GiRs, and the mechanical models and design proposals for glued-in rods, are given in [11]. A major part of the fundamental work has been done in previous research projects, e.g. the development of mechanical models, studies of softening behaviour, studies on the impact of moisture variation and duration of load, and the evaluation of different adhesive types in [12–17], COST E 34 [18] and COST FP 1004 [19].

Most of these studies focus on the use of GiR as connecting elements, but the current study is focused on the use of GiR as reinforcing elements. One of the major drawbacks of a wider application of GiR in practice is the lack of standardized design procedures. Some important aspects for the standardization of GiR as reinforcement are summarized in the next sections.

2 Properties of Glued-in Rods

2.1 Geometrical Properties

The components of a glued-in rod system are depicted in Fig. 1, namely the rod, the adhesive and the surrounding timber. The most important geometrical parameters are the cross-sectional area of the wood $A_w = b \cdot h$, the area of the rod A_r (depending on the model used for strength calculation, A_r can be based, for example, on the effective diameter of a threaded rod d_r, which in turn is based on its own nominal diameter, or on the nominal diameter of a rebar rod), the thickness of the bondline t, and the anchorage (bonded) length of the rod l_a. The drill hole is characterized by its length l_{drill} and diameter d_{drill}. The bondline may be shifted by the not bonded length l_{nb} from the wood surface. The load direction in relation to the rod in axial direction

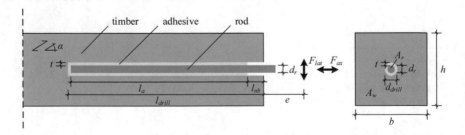

Fig. 1 Parts and geometrical parameters of a glued-in rod

Table 1 Stiffness properties of selected solid timber and glued laminated timber strength classes according to [23, 24]

Material	Strength class	$E_{0,mean}$ (N/mm^2)	$E_{90,mean}$ (N/mm^2)
Solid timber	C24	11,000	370
	D40	13,000	860
Glued laminated timber	GL24h	11,500	300

F_{ax} (e.g. pull-out) or lateral direction F_{lat} at distance e from the wood surface and the rod to grain angle α are, of course, also of major importance.

2.2 Material Properties

A GiR is a highly complex composite system consisting of the materials of the rod, adhesive and the surrounding timber. The properties of the individual materials have a considerable impact on the overall behaviour of the system. In the following, loading in the axial direction of the rod is assumed if no direction is stated.

2.2.1 Timber

The stiffness of the timber in the direction of the rod axis has a major impact on the stress distribution along the bondline. Hence, it is important to take into account the stiffness parallel to the grain E_0, for rods installed along the grain, and the stiffness perpendicular to the grain E_{90}, for rods installed perpendicular to the grain, as summarized in Table 1. Strength parameters such as shear and tensile strengths, are important to account for failure in the timber other than the bondline failure. Laminated-veneer lumber (LVL) with crossband veneers allows for a redistribution of stresses between single layers and higher load-carrying capacities can be observed [20–22]. For engineered wood products with cross layers such as CLT, experience is limited but similar behaviour can be expected.

2.2.2 Adhesives

The adhesive is required for the transfer of forces between the rod and the timber. Failure can occur not only in the adhesive itself but also in the interface between the rod and adhesive or between the timber and adhesive. This is why the choice of an appropriate adhesive is of major importance for the performance of a GiR. Threaded or deformed rods reduce the risk of cohesive failure between rod and adhesive by mechanical interlocking.

Table 2 Material properties of phenol-resorcinol (PRF), polyurethane (PUR) and epoxy resin (EPX) adhesives. The fracture energy refers to the effective fracture energy as described in [14]

Material	References	τ_f (N/mm^2)	G_f (N/mm^2)
PRF	[27]	8.9	4.15
	[14]	6.6	2.90
PUR	[27]	9.7	1.77
	[14]	10.6	0.91
EPX	[27]	10.5	1.89
	[14]	12.1	1.31

The stress distribution along the bondline and the failure behaviour of the bondline mainly depend on the shear stiffness G/t, where G is the shear modulus of the adhesive and t is the thickness of the bondline, the shear strength τ_f and the fracture energy G_f of the bondline. With regard to G_f the terminology from non-linear fracture mechanics is used [25]. This theory can be seen as a general framework including linear elastic fracture mechanics and elasto-plasticity as special cases as discussed in Sect. 3.

The most commonly used adhesives for GiRs are polyurethane (PUR) and epoxy resins (EPX). The advantages of these adhesives are amongst others good gap filling properties. To some extent also phenol resorcinol (PRF) has been used. The advantage of this adhesive is its good thermal stability and resistance against high moisture contents. Due to its inadequate gap filling properties and its insufficient bond with steel rods it is commonly not used for GiR anymore. These adhesives exhibit partly considerable differences in their mechanical properties and their handling, as discussed in [26]. In Table 2, a summary of material property values from the literature [14, 27] are given. Values obtained from [14] are based on small-scale tests and those obtained from [27] are based on backward calculation (fitting) of tests on large-scale specimens. Detailed information on both sets of data is also available in [28].

2.2.3 Rod

Steel is the most common rod material for GiR. It offers the potential to create well defined plastic deformations in the connections in order to enable load-redistribution between single rods within a connection [29]. Rods with a metric thread but also rebar rods are used in GiRs, the latter especially in combination with concrete as for timber-concrete-composite elements. The use of smooth rods is not recommended due to the insufficient mechanical interaction between the rods and adhesive. For reinforcement purposes or when non-metallic connections are necessary [30] also fibre-reinforced polymers (FRP) such as glass or carbon FRP are used (GFRP or CFRP, respectively). The axial stiffness of the rod has an important impact on the stress distribution along the bondline. The considerable differences in the stiffness of rods of different materials, as summarized in Table 3, require different choices of the relative dimensions of the GiR to achieve optimal structural behaviour.

Material	Stiffness (N/mm²)
Steel	210,000
Glass fibre	35,000–51,000
Carbon fibre	120,000–580,000

Table 3 Stiffness properties of rods according to [31, 32]

3 Mechanical Models

Different models are used to describe the mechanical behaviour of the system and the stress distribution around GiR. The different theories used are traditional strength analysis (i.e. determining the load-bearing capacity based on a stress or strain criterion), linear elastic fracture mechanics, and non-linear fracture mechanics [25]. The notations shown in Fig. 2 are used to describe the mechanical models: E_1 and the cross-sectional area A_1 of the rod, E_2 and the cross-sectional area A_2 of the timber, the shear modulus G_3 and the thickness t_3 of the adhesive layer, radius r from the centre of the rod to the interface between adhesive and timber, the position x along the rod of length l, the applied load P, the distributed load in the timber Q, and the normal force N to the cross-section.

3.1 Strength Theory

A number of different GiR models are discussed below. These are analytical and one-dimensional in the sense that the adherents (rod and timber) act in pure tension/compression and the bondline in pure shear. A first approximation of the impact of the three-dimensional structure of a GiR can be achieved by a Thimoshenko

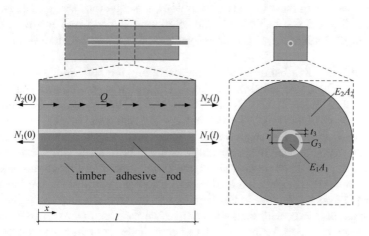

Fig. 2 Geometrical parameters and material properties of a GiR

beam model, as suggested by [28]. This model can be used, for example, to determine the tensile stresses along the bondline. The non-linear degradation and mixed mode failure conditions can be accounted for by the finite element model of the GiR, as discussed in [33].

3.1.1 Volkersen Approach

A description of the shear stress distribution in lap joints was presented by Volkersen [34, 35]. The application of this so-called Volkersen model for GiRs is discussed in detail in [28]. Assuming only normal strain in the rod and wood materials and pure shear in the adhesive layer, the following differential equation for the shear stress, $\tau(x)$, can be stated:

$$\tau'' - \omega^2 \tau = \frac{-QG_3}{t_3 E_2} \quad \text{with} \quad \omega^2 = \left(\frac{G_3 2\pi r}{t_3}\right)\left[\frac{1}{A_1 E_1} + \frac{1}{A_2 E_2}\right] \tag{1}$$

The general solution for this differential equation is:

$$\tau = C_1 \cosh(\omega x) + C_2 \sinh(\omega x) + \tau_P \tag{2}$$

The term τ_P depends on the distribution of the shear load in the timber and is $\tau_P = 0$ for $Q(x) = 0$. For a constant distributed load, Q_0, we obtain:

$$\tau_P = \frac{Q_0 G_3}{t_3 E_2 \omega^2} \tag{3}$$

The different applications of a GiR with specific loading conditions, as summarized in Table 4, can be solved by assigning corresponding boundary conditions at $x = 0$ and $x = l$. With these boundary conditions, the equations for shear stress distributions in Table 5 can be derived. The solutions for load cases 1–5 were derived already in [28] while the solution for load case 6 is a novel feature of the present work.

The different distribution of shear stresses along the bondline, according to the equations in Table 5, is illustrated in Fig. 3. The following parameters were used for this illustration: $E_1 = 210{,}000$ N/mm^2, $E_2 = 11{,}500$ N/mm^2, $G_3 = 16$ N/mm^2, $A_1 = 200$ mm^2, $A_2 = 10{,}000$ mm^2, $t_3 = 0.5$ mm, $r = 8.5$ mm, and $l = 1000$ mm. A force $P = 10$ kN was applied. The impact of the different load cases and boundary conditions is obvious from the analytical derivations and can be confirmed by Finite Element Analysis [28, 36]. Tests on GiRs in connections have been carried out mostly for pull-pull and pull-compression load cases. A summary of tests in the literature is given in [11]. As can be seen in Fig. 3, these two load cases both provoke the highest stress peaks at $x = 0$, with the pull-compression case leading to an approximately 25% higher stress level for the investigated combination of material and geometrical parameters.

Table 4 Load cases and boundary conditions

Load case		Example	Illustration
1	Pull-pull	Tensile connection	
2	Pull-compression	Test situation	
3	Pull of rod	Increased temperature in metal rod	
4	Pull of wood	Increased MC in wood	
5	Pull-distributed	Tensile connection perp. to grain	
6	Compression-distributed	Compression reinforcement for concentrated load	

Table 5 Shear stress distributions for different load cases

Load case		
1	Pull-pull (lap joint)	$\tau = \frac{PG_3}{t_3\omega E_1 A_1}\left\{\left[\cosh(\omega l) + \frac{E_1 A_1}{E_2 A_2}\right]\frac{\cosh(\omega x)}{\sinh(\omega l)} - \sinh(\omega x)\right\}$
2	Pull-compression	$\tau = \frac{PG_3}{t_3\omega E_1 A_1}\left[1 + \frac{E_1 A_1}{E_2 A_2}\right]\left\{\frac{\cosh(\omega x)}{\tanh(\omega l)} - \sinh(\omega x)\right\}$
3	Pull of rod (increased temperature)	$\tau = \frac{PG_3}{t_3\omega E_1 A_1}\left\{[\cosh(\omega l) - 1]\frac{\cosh(\omega x)}{\sinh(\omega l)} - \sinh(\omega x)\right\}$
4	Pull of wood (increased MC)	$\tau = -\frac{PG_3}{t_3\omega E_1 A_1}\frac{E_1 A_1}{E_2 A_2}\left\{[\cosh(\omega l) - 1]\frac{\cosh(\omega x)}{\sinh(\omega l)} - \sinh(\omega x)\right\}$
5	Pull-distributed	$\tau = \frac{PG_3}{t_3\omega E_1 A_1}\left\{\cosh(\omega l)\frac{\cosh(\omega x)}{\sinh(\omega l)} - \sinh(\omega x) + \frac{E_1 A_1}{E_2 A_2\omega l}\right\}$
6	Compression-distributed	$\tau = \frac{PG_3}{t_3\omega E_1 A_1}\frac{E_1 A_1}{E_2 A_2\omega l}\{\tanh(\omega l)\sinh(\omega x) - \cosh(\omega x) + 1\}$

Note that for cases involving a distributed load Q_0, $P = Q_0/(l \cdot A_2)$

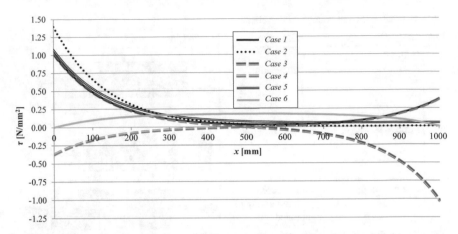

Fig. 3 Stress distribution along the bondline for different loading cases according to the linear elastic theory

Load cases 3 (pull of rod), 4 (pull of wood), 5 (distributed-pull) and 6 (compression-distributed) are all important for reinforcement purposes. In Fig. 4 the relative deformation of the timber surrounding the rods due to changes in MC of the wood and due fracture induced by tension perpendicular to grain stresses are sketched. The shrinkage of the timber for a decrease in MC as shown in Fig. 4a is constrained by the GIR causing tension perpendicular to the grain stresses in the timber and increasing the risk of cracking. In case of an increase in MC as shown in Fig. 4b the rod is loaded in tension. In areas loaded tension perpendicular to the grain such as curved beams, reinforcement may be used to avoid separation and to enable load transfer between cracked members.

Load case 4 represents the situation of a GiR loaded by the shrinkage of the timber due to decreasing MC. GiR in applications as reinforcement are often exposed to this loading situation. The rod to grain angle in these applications can vary between directions parallel and perpendicular to the grain. In Fig. 5 the impact of the different stiffness of the timber parallel and perpendicular to the grain on the shear stress distribution along the bondline is illustrated. In addition to the above stated parameters, the following parameters were used for the illustration: $E_{2,0} = 11{,}500$ N/mm^2 and

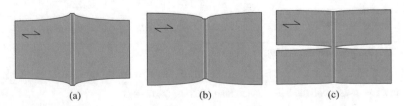

Fig. 4 Examples of reinforcement perpendicular to the grain exposed to decreased MC (**a**), increased MC (**b**) and a crack due to tension perpendicular to the grain (**c**)

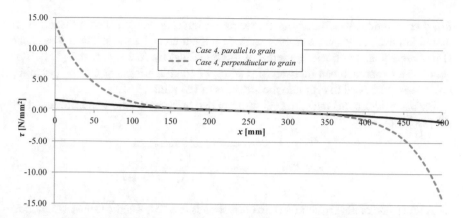

Fig. 5 Effects of different rod to grain directions on the shear stress distribution along the bondline for load case 4 (pull of wood)

$E_{2,90} = 350$ N/mm^2. The force in the GiR, $P = \Delta MC \, \alpha \, E_2 A_2$, was calculated for a change in moisture content $\Delta MC = 4\%$, and hygroexpansion factors $\alpha_{90} = 0.25$ and $\alpha_0 = 0.01$, respectively.

The failure load of a GiR can be calculated according to traditional strength theory: at the position of the highest stress peak according to the stress equations in Table 5, the stress is assumed to be equal to the failure stress, i.e. $\tau = \tau_f$. For load cases 1, 2 and 5, the highest stresses occur at $x = 0$ for $E_1 A_1 / E_2 A_2 < 1.0$. This assumption of brittle failure neglects the possibility of stress redistribution due to, for example, fracture softening or plasticity in the post-failure behaviour.

3.1.2 Generalised Volkersen Approach

One problem that arises when applying traditional strength theory to the equations in Table 5 is that the stiffness of the adhesive layer G_3/t_3 might be hard to determine or unknown. Another problem is that a traditional linear elastic approach will, in general, underestimate the load-bearing capacity, since non-linear effects due to plasticity or crack initiation and propagation are not taken into account. By exchanging the elastic stiffness of the bondline with a measure of stiffness derived from the bondline shear strength and fracture energy, Gustafsson [37], derived the so-called generalised Volkersen approach, which assumes that:

$$\frac{G_3}{t_3} = \frac{\tau_f^2}{2 G_f} \tag{4}$$

The approach expressed in Eq. 4 can be justified by the fact that for cases where the *shape* of non-linear softening bondline behaviour has no effect on the load-bearing capacity of the joint, *any* such shape can be assumed. The shape assumed when using

Eq. 4 corresponds to a linear elastic response, but with the stiffness adjusted such that at failure, i.e. when the stress reaches τ_f, the area under the stress versus shear slip curve equals the fracture energy of the bondline, G_f (see Fig. 6 and [37]). The generalised approach can be used to predict the load-bearing capacity of the joint, but it cannot be used to estimate the stiffness of the joint.

By inserting Eq. (4) into Eq. (1), the so-called brittleness ratio was introduced by [37] as follows:

$$\varpi = \omega l = \sqrt{\frac{\tau_f^2 \, l^2 \pi \, r}{G_f}} \sqrt{\frac{1}{E_1 A_1} + \frac{1}{E_2 A_2}} \qquad (5)$$

A low value of the brittleness ratio corresponds to a very ductile failure behaviour. For $\varpi = 0$, a perfectly plastic load deformation behaviour exists, and no failure for the load cases 3 and 4 with only internal stresses can occur. The failure load of the GiR is dependent on this brittleness ratio, as summarized in Table 6.

Fig. 6 Example of the non-linear shear stress-shear slip curve along the bondline of an adhesive with fracture energy G_f and simplified linear shear stress-shear slip behaviour with equal fracture energy

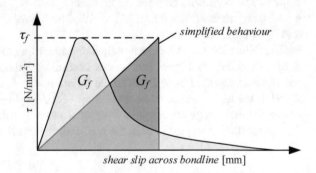

Table 6 Maximum loads for different loading cases

Load case		
1	Pull-pull (lap joint)	$\frac{P_f}{2\pi r l \tau_f} = \frac{1}{\omega}\left(1 + \frac{E_1 A_1}{E_2 A_2}\right)\left[\frac{\sinh(\omega l)}{\cosh(\omega l) + \frac{E_1 A_1}{E_2 A_2}}\right]$
2	Pull-compression	$\frac{P_f}{2\pi r l \tau_f} = \frac{1}{\omega}[\tanh(\omega l)]$
3	Pull of rod (decreased MC)	$\frac{P_f}{2\pi r l \tau_f} = \frac{1}{\omega}\left(1 + \frac{E_1 A_1}{E_2 A_2}\right)\left[\frac{\sinh(\omega l)}{\cosh(\omega l) - 1}\right]$
4	Pull of wood (increased MC)	$\frac{P_f}{2\pi r l \tau_f} = \frac{1}{\omega}\left(1 + \frac{E_2 A_2}{E_1 A_1}\right)\left[\frac{\sinh(\omega l)}{\cosh(\omega l) - 1}\right]$
5	Pull-distributed	$\frac{P_f}{2\pi r l \tau_f} = \frac{1}{\omega}\left(1 + \frac{E_1 A_1}{E_2 A_2}\right)\Big/\left[\frac{\cosh(\omega l)}{\sinh(\omega l)} + \frac{E_1 A_1}{E_2 A_2 \omega l}\right]$
6	Compression-distributed[a]	$\frac{P_f}{2\pi r l \tau_f} = \left(1 + \frac{E_2 A_2}{E_1 A_1}\right)\Big/\left[1 - \frac{1}{\cosh(\omega l)}\right]$

[a]Note that the load for this case is transferred partly by shear stress along the bondline and partly by contact stress in the wood (see Table 4)

3.2 Linear Elastic Fracture Mechanics Theory

To derive a linear elastic fracture mechanic (LEFM) solution of the load-bearing capacity of the GiR, a small extension of the fracture process zone compared to the length of the joint is assumed. This LEFM solution can be derived from the change in compliance, dC, of the GiR joint due to an incremental increase in the fracture zone from $x = 0$ to $x = dx$ for the load cases pull-pull, pull of rod and pull-distributed, as presented in [37].

$$dC = \frac{dx}{E_1 A_1} - \frac{dx}{E_1 A_1 + E_2 A_2} \tag{6}$$

With the change in potential energy:

$$dW = -\frac{1}{2} P^2 dC \tag{7}$$

The load at progression of the fracture zone can be calculated when applying the energy release rate $-dW = G_f\, 2\pi r\, dx$ as follows

$$P_{max} = 2\sqrt{\pi r E_1 A_1 G_f} \sqrt{1 + \frac{E_1 A_1}{E_2 A_2}} \tag{8}$$

For the load case 2 (pull-compression), the crack propagation load is:

$$P_{max} = 2\sqrt{\pi r E_1 A_1 G_f} \left/ \sqrt{1 + \frac{E_1 A_1}{E_2 A_2}} \right. \tag{9}$$

For the load case 4 (pull of wood) as being relevant for swelling of the timber due to an increase in MC, the crack propagation load is:

$$P_{max} = 2\sqrt{\pi r E_1 A_1 G_f} \sqrt{\left(\frac{E_2 A_2}{E_1 A_1}\right)^2 + \frac{E_2 A_2}{E_1 A_1}} \tag{10}$$

The LEFM model does not account for the bonded length of the rod. It is noticeable that the above LEFM equations can also be obtained by letting the joint brittleness approach infinity in the equations of Table 6.

3.3 Non-linear Elastic Fracture Mechanics Theory

The post-failure behaviour of the adhesive is accounted for by the non-linear fracture mechanics (NLFM) theory. In [37, 38] a lap joint is analysed by describing the failure behaviour by a bilinear behaviour with an elastic stiffness G_3 and a post failure stiffness G_3^*. The transition between pre- and post-failure regimes is defined by the failure point with the failure stress τ_f at a deformation δ_f, as illustrated in Fig. 7. In [38] the case of a symmetric lap joint with $E_1 A_1 = E_2 A_2$ is presented. In this section, the general solution for a GiR as reinforcement with an arbitrary diameter for load cases 3 and 4, according to Table 4, is discussed. This solution benefits from the symmetric stress distribution along the bondline for these load cases if the coordinate system is defined as in Fig. 7.

It can be distinguished between hardening with $G_3^* > 0$, ideal plasticity with $G_3^* = 0$, and softening with $G_3^* < 0$, as illustrated in Fig. 8. The limit case, $G_3^* \to -\infty$, describes perfectly brittle failure behaviour according to traditional strength theory.

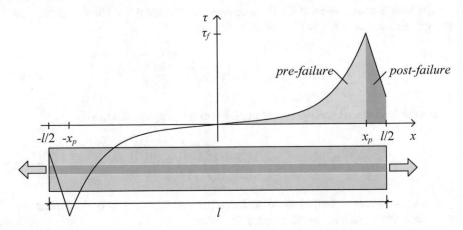

Fig. 7 Illustration of the distribution of shear stresses along the bondline of a GiR with pre-failure and post-failure zones

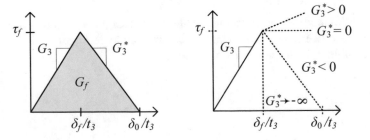

Fig. 8 Examples of simplified stress-deformation behaviour of adhesives

For simplicity, it has been assumed that the elastic stiffness G_3 is the same for the NLFM theory and the generalised Volkersen theory. Differences between these theories could be evaluated in a more detailed NLFM study using the two shear stress-shear slip behaviours illustrated in Fig. 6.

The stress distribution along the bondline can be described by separating the stresses into a linear elastic pre-failure region and a post-failure region, as shown in Fig. 7. Within the linear elastic pre-failure zone between $0 \leq x \leq x_p$, the elastic stresses τ_{el} can be derived from the differential Eq. (1). From the symmetry of the shear stress distribution it follows that $C_1 = 0$ in Eq. (2). Together with the boundary condition $\tau_{el}(x_p) = \tau_f$ at the instant of failure and with $Q = 0$ for load cases 3 and 4, the shear stresses can be described as follows:

$$\tau_{el} = \frac{\tau_f \sinh(\omega x)}{\sinh(\omega x_p)} \tag{11}$$

The stiffness of the post-failure zone, $x_p \leq x \leq l/2$, is negative for the case of softening behaviour ($G_3^* < 0$). This leads to a different solution of Eq. (1) as follows:

$$\tau_{pl} = C_3 \sin \omega^* x + C_4 \cos \omega^* x \quad \text{with} \quad \omega^* = \sqrt{-\left(\frac{G_3^* 2\pi r}{t_3}\right)\left[\frac{1}{A_1 E_1} + \frac{1}{A_2 E_2}\right]} \tag{12}$$

The constants C_3 and C_4 can be determined from the boundary conditions $\tau_{pl}(x_p)$ $= \tau_f$ and $N_1(l/2) = P$. The failure load, P_f, at which the failure region progresses can be derived from the transitional condition between the pre- and post-failure zones at $x = x_p$:

$$\left(\frac{1}{G_3^*}\frac{\delta \tau_{pl}}{\delta x}\right) = \left(\frac{1}{G_3}\frac{\delta \tau_{el}}{\delta x}\right) \tag{13}$$

The coupling of the equations of pre- and post-failure stresses leads to a somewhat lengthy expression of P_f. The maximum of P_f along the progress of x_p corresponds to the load-carrying capacity, $P_{f,max}$, of the GiR. The load-carrying capacity can be calculated using $P_{f,max} = P_f(x_p)$ for x_p such that $\delta P_f/\delta x_p = 0$. However, a general analytical solution is hard to find and solutions can instead preferably be found by numerical integration.

The model is valid only before full separation of the adhesives has occurred, i.e. as long as $\delta \leq \delta_0$ (cf. Fig. 8) at $x = l/2$ (i.e. for $\tau_p(l/2) \geq 0$). This situation is shown in Fig. 9a with a progressive increase in the failure region with decreasing x_p. The equation for the stress distribution is valid in the range $180 \text{ mm} \leq x_p \leq 250 \text{ mm} = l/2$. In Fig. 9b the corresponding development of the load, P_f, at which the failure region progresses is illustrated. The load-carrying capacity for $x_p = 193$ mm is $P_{f,max}/(ld\pi)$ $= 3.4 \text{ N/mm}^2$. It should be noted that the analysis of additional crack propagation is equivalent to performing the above analysis but assuming a shorter glued-in length.

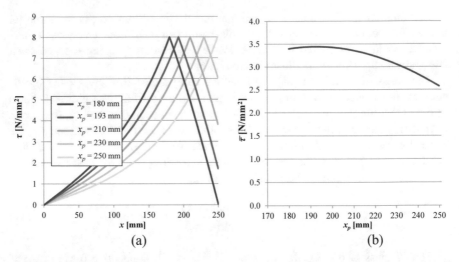

Fig. 9 Shear stress distribution along the bondline (**a**) and mean shear stress $\bar{\tau} = P/(ld\pi)$ (**b**) as functions of the position of the stress peak (x_p)

The curves of the shear stress distribution along the bondline in Fig. 9 have been derived with the following parameters for a GiR perpendicular to the grain: $E_1 = 210000$ N/mm², $E_2 = 350$ N/mm², $A_1 = 201$ mm², $A_2 = 62500$ mm², $t_3 = 0.5$ mm and $l = 500$ mm. The shear stiffness properties of the bondline are $G_3 = 16$ N/mm² and $G_3^* = -16$ N/mm², and the shear strength at failure initiation is $\tau_f = 8$ N/mm². From these properties, the fracture energy can be determined to be $G_f = 2$ N/mm, according to Eq. (14).

$$G_f = \tau_f^2 t_3 \left(\frac{G_3^* - G_3}{2G_3^* G_3} \right) \tag{14}$$

With the three parameters τ_f, G_3 and G_3^* or with G_f, respectively, it is possible to describe the full stress-separation behaviour of the adhesive in a simplified way. The case of perfectly brittle failure, according to strength theory, can be represented by inserting $G_3^* = -\infty$ N/mm² into Eq. (14), which leads, for the above stated material properties, to a fracture energy $G_f = 1$ N/mm.

The governance of either LEFM, NLFM or plasticity theory can be evaluated by plotting the normalized strength $\bar{\tau}/\tau_f$ against the square of the brittleness ratio ϖ, as shown in Fig. 10. For large values of ϖ^2, the brittle failure in the bondline can be described by LEFM and the normalized strength decreases with an inclination of $-1/2$. The transition between LEFM and plasticity theory described by NLFM depends on the geometry and the material properties of wood, rod and adhesive, as shown in Fig. 10. In contrast to a GiR in pull-pull loading conditions, the normalized strength does not approach 1.0 for small brittleness ratios due to the different loading

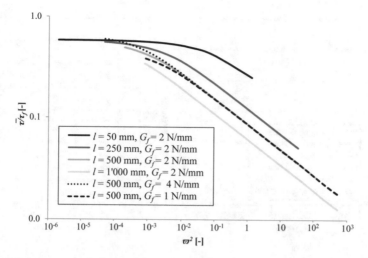

Fig. 10 Normalized shear strength $\bar{\tau} / \tau_f$ plotted against the square of brittleness ratio ϖ^2

conditions. In order to reach the highest possible load-carrying capacity of a GiR of any given length, the brittleness ratio ϖ should be as small as possible.

4 Factors that Affect GiR Performance

When planning to use GiR in connections or as reinforcement, certain boundary conditions and parameters are often predefined. In the case of reinforcement for tension perpendicular to the grain of a beam, the required bonded length should be given. An adjustment of the performance of the GiR is possible only with regard to its relative diameter $E_1A_1/(E_2A_2)$, its material (which affects the stiffness E_1), and the adhesive used. In the following section, the impact of some of these properties is evaluated.

4.1 Strength and Softening Behaviour of Adhesive

The softening behaviour of adhesives allows stress redistribution during progressive failure. With decreasing steepness of stiffness in the post-failure regime, the failure behaviour becomes more and more plastic. In Fig. 11a, the impact of different values of stiffness G_3^* and the corresponding fracture energies G_f is shown for the same configuration as in Fig. 9. For the very steep decrease, $G_3^* = -256\,\mathrm{N/mm^2}$, brittle failure occurs immediately after failure initiation. In contrast, for $G_3^* = -2\,\mathrm{N/mm^2}$

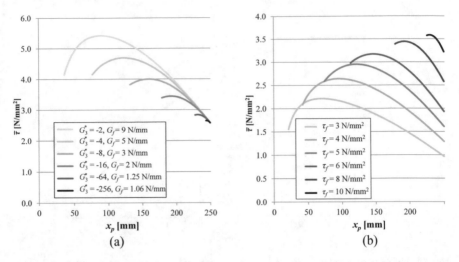

Fig. 11 Impact of softening behaviour (**a**) and shear strength of adhesive (**b**) on the mean shear stress $\bar{\tau} = P/(ld\pi)$ at failure

with a higher fracture energy value, further load increase is possible with increasing x_p.

The ultimate shear strength τ_f determines the load at failure initiation. In Fig. 11b, the impact of different shear strength values is shown when a constant initial stiffness $G_3 = 16$ N/mm^2 and a constant fracture energy $G_f = 2$ N/mm are assumed for all curves. It can be seen that the load-carrying capacity is only slightly dependent on τ_f. Hence, the shear strength value has an impact on the load only at failure initiation.

4.2   Relative Size and Stiffness of Rod

To achieve a smooth load transfer and avoid high stress peaks, the ratio E_1A_1/E_2A_2 between rod and wood should be close to 1 [19]. The stiffness of a rod can differ considerably across different rod materials, such as GFRP, CFRP and steel, as summarized in Table 3. When the rod diameters are known, the relative area of the timber can be adjusted by choosing the optimal spacing between the GiRs and edge distances. In Fig. 12, the impact of the spacing a_1 along the grain direction between GiRs, used as tension perpendicular to grain reinforcement in a beam, on the load-carrying capacity of the GiRs is shown. It is here assumed that the effective wood area that can be activated in the joint is linearly dependent of the inter rod distance, which in reality, however, is limited. The curves are given for different rod diameters, rod materials and fracture energies. The dimensions of the beam are $h = 500$ mm and $b = 140$ mm. The case, $E_1A_1/E_2A_2 = 1$, is met for $a_1 = 861$ mm for steel rods with $d = 16$ mm. Choosing an adhesive with a higher fracture energy has a similar effect as choosing a large rod diameter.

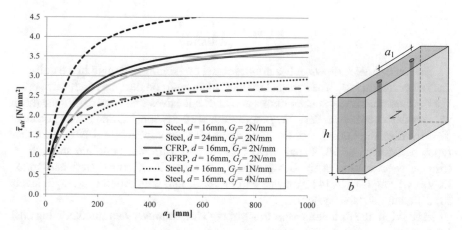

Fig. 12 Impact of the distance a_1 between GiRs on the mean shear capacity $\bar{\tau}_{ult} = P_{ult}/(hd\pi)$

5 Design of GiR

The design of GiRs should be simple and transparent. So far, no generally accepted design procedure has been established in European design codes, but the specification of design clauses is given as a task for the revision of Eurocode 5 [39]. Activities with regard to the specification of test standards for bondline strength for GiR are ongoing in technical committees of CEN and ISO. When specifying design clauses, it should not be focussed only on the bondline strength but also on timber failure modes and the tensile failure of the rod. In general, it should always be aimed at achieving a ductile failure of the steel rod instead of brittle failure in the surrounding timber or in the bondline. Overstrength of the steel and a reliable estimation of bondline strength are challenges for the development of design equations. A comprehensive summary of various design equations is given in [11]. Most design proposals for GiR can be expressed in the form of the following equation:

$$F_{ult} = \pi d \, l \, f \, k \tag{15}$$

In this design equation, d is a diameter, e.g. the borehole diameter, thread diameter or an equivalent diameter of the rod, and f is a strength parameter of the bondline. This strength parameter is often associated with the shear strength of the adhesive τ_f and also accounts for the softening behaviour of the adhesive, i.e. the fracture energy. In addition, the strength parameter can be reduced in dependency of the diameter of the rod or borehole, length of the rod, and other parameters. Recent evaluations of a large amount of test data show that within the range of geometries with relevance for practice and for adhesives currently used the impact of diameter on the strength bondline can be neglected. A power law specification of the bondline strength solely in dependency of the bondline length may be sufficient.

$$f = const. \left(\frac{l_{ad}}{250\text{mm}} \right)^{0.5} \tag{16}$$

The rod diameter shows only a minor impact on bondline strength in the pull-pull tests evaluated in [40]. The constant in Eq. (16) depends on the type of adhesive used. For the large number of special Epoxy and PUR adhesives evaluated in [40] the mean value of the constant was derived to approximately $6.5\,\text{N/mm}^2$. On characteristic level (5%-fractile) the constant in Eq. (16) amounts approximately $4.5\,\text{N/mm}^2$. Other types of Epoxy and PUR may yield considerably different bondline strength. A trilinear specification of the bondline strength (as in the German National Annex DIN EN 1995-1-1/NA [41]) may be chosen for design with different strength levels for different adhesive systems.

In addition, the following aspects are of high importance when standardizing GIR in connections or as reinforcement:

- compatibility of adhesive, rod and timber as discussed in [12, 27];
- quality control [42];
- measures for reducing stress singularities [43];
- impact of variation in the moisture content and of duration of load effects [44, 45];
- required distances between rods and edge distances [6];
- fatigue and the behaviour under cyclic loading [46, 47].

6 Conclusions and Outlook

The GiR in connections or as reinforcement is a highly complex system of different materials which requires adequate mechanical models for the determination of its stress distribution and load-carrying capacity. The failure behaviour of the adhesive has an important impact on the failure behaviour of the GiR.

As discussed in this chapter, the load-carrying capacity of a GiR is highly dependent on the shear strength and softening behaviour of the adhesive. Hence it is crucial to account for the adhesive properties when designing a GiR. This can either be done by explicitly including the adhesive properties in the design equation or by defining certain material-dependent constants for the design approach for each different class of adhesive. Nevertheless, research on adequate testing procedures for the determination of the relevant material properties of adhesives in GiR application is essential and some work has already been done [44, 48]. In addition, it would be of great interest to further develop the design expressions such that common situations when GiRs are used for reinforcement can be analysed. For GiRs in general (both for joints and for reinforcement), further theoretical and experimental studies on the local behaviour and failure of the wood close to the rod would be of interest. The goal would be to find a rational basis for design rules regarding distances between rods and the distance from rod to edge. For GiRs in joints, there is an evident lack of experimental and theoretical investigations on rods loaded laterally and under combined axial and lateral load.

References

1. Baumeister A et al (1972) Versuche für neuere Großbauten im Ingenieurholzbau. Bauen mit Holz 74(6):314–317
2. Edlund G (1975) In Brettschichtholz eingeleimte Gewindestangen. Svenska Trgforskningsinstitutet, Med. Serie B, Nr.33, Stockholm
3. Riberholt H (1978) Eingeleimte Gewindestangen. Danmarks Tekniske Höjskole, Lyngby. Rapport Nr. R 99
4. Steiger R, Gehri E, Widmann R (2007) Pull-out strength of axially loaded steel rods bonded in glulam parallel to the grain. Mater Struct 40(1):69–78. https://doi.org/10.1617/s11527-006-9111-2
5. Widmann R, Steiger R, Gehri E (2007) Pull-out strength of axially loaded steel rods bonded in glulam perpendicular to the grain. Mater Struct 40(8):827–838. https://doi.org/10.1617/s11527-006-9214-9
6. Blaß HJ, Laskewitz B (2001) Glued-in rods for timber structures—effect of distance between rods and between rods and timber edge on the axial strength. Versuchsanstalt für Stahl, Holz und Steine, Abteilung Ingenieurholzbau, Universität Fridericiana, Karlsruhe, Germany
7. Gattesco N, Gubana A (2006) Performance of glued-in joints of timber members. In: Proceedings of the 9th world conference on timber engineering, pp 1848–1855
8. Gattesco N, Gubana A (2001) Experimental tests on glued joints under axial forces and bending moments. In: Joints in timber structures, international RILEM symposium, Stuttgart, Germany, pp 353–362
9. Vašek M, Vyhnálek R (2006) Timber semi rigid frame with glued-in-rods joints. In: Proceedings of the 9th world conference on timber engineering Portland, Oregon, pp 1825–1832
10. Broughton JG, Hutchinson AR (2001) Pull-out behaviour of steel rods bonded into timber. Mat Struct 34(2):100. https://doi.org/10.1007/BF02481558
11. Tlustochowicz G, Serrano E, Steiger R (2011) State-of-the-art review on timber connections with glued-in steel rods. Mater Struct 44(5):997–1020. https://doi.org/10.1617/s11527-010-9682-9
12. Kemmsies M (1999) Comparison of pull-out strengths of 12 adhesives for glued-in rods for timber structures. Technical report, SP Swedish National Testing and Research Institute, Borås, Sweden. SP Report 1999:20
13. Aicher S, Gustafsson PJ, Wolf M (1999) Load displacement and bond strength of glued-in rods in timber influenced by adhesive, wood density, rod slenderness and diameter. In: Boström L (ed) Proceedings 1st RILEM symposium on timber engineering, Stockholm, September 13–14, 1999. PRO 8. RILEM Publications, France. ISBN: 2-912143-10-1
14. Serrano E (2001) Glued-in rods for timber structures—an experimental study of softening behaviour. Mater Struct 34:228–334. https://doi.org/10.1007/BF02480593
15. Aicher S (2001) Characteristic axial resistance of threaded rods glued-in spruce dependent on adhesive type—a complementary database for the GIROD project. Otto-Graf-Institut, Stuttgart, Germany
16. Aicher S (2001) Glued-in rods for timber structures (GIROD). WP 4—effect of moisture conditions. Technical Report. Otto-Graf-Institut, Stuttgart, Germany
17. Aicher S (2001) Glued-in rods for timber structures (GIROD). WP 5—duration of load tests on full-sized glued-in rod specimens. Technical Report, Otto-Graf-Institute, Stuttgart, Germany
18. Serrano E, Steiger R, Lavisci P (2008) Glued-in rods. In: Core document of the COST action E34 bonding of timber, Vienna, Austria. University of Natural Resources and Applied Life Sciences

19. Steiger R, Serrano E, Stepinac M, Rajčić V, O'Neill C, McPolin D, Widmann R (2015) Strengthening of timber structures with glued-in rods. Constr Build Mater 97:90–105. https://doi.org/10.1016/j.conbuildmat.2015.03.097

20. Hunger F, Stepinac M, Rajčić V, van de Kuilen J-WG (2016) Pull-compression tests on glued-in metric thread rods parallel to grain in glulam and laminated veneer lumber of different timber species. Eur J Wood Wood Prod 74(3):379–391. https://doi.org/10.1007/s00107-015-1001-2

21. Stepinac M, Rajčić V, Hunger F, Van de Kuilen JWG (2016) Glued-in rods in beech laminated veneer lumber. Eur J Wood Wood Prod 74(3):463–466. https://doi.org/10.1007/s00107-016-1037-y

22. Meyer N, Blaß HJ (2017) Fachwerkträger aus Buchenfurnierschichtholz. Bautechnik 94(11):751–759. https://doi.org/10.1002/bate.201700071

23. EN 338 (2009) Structural timber—strength classes. European Committee for Standardization CEN, Bruxelles, Belgium

24. EN 14080 (2013) Timber structures—glued laminated timber and glued solid timber—requirement. European Committee for Standardization CEN, Bruxelles, Belgium

25. Serrano E, Gustafsson PJ (2007) Fracture mechanics in timber engineering, strength analyses of components and joints. Mater Struct 40(1):87–96. https://doi.org/10.1617/s11527-006-9121-0

26. Aicher S (2003) Structural adhesive joints including glued-in bolts. In: Thelandersson S (ed) Timber engineering. Wiley, Chichester

27. Johansson CJ, Pizzi T, Van Leemput M (2002) State of the art—report, COST Action E13, Wood Adhesion and Glued Products, Working Group 2: Glued Wood Products. ISBN 92-894-4892-X

28. Gustafsson PJ, Serrano E (2001) Glued-in rods for timber structures—development of a calculation model. Report TVSM-3056. Lund University, Division of Structural Mechanics, Lund

29. Gehri E (2016) Performant connections—a must for veneer-based products. In: Proceedings of WCTE 2016—world conference on timber engineering, Vienna, Austria

30. Ansell MP, Smedley D (2007) Briefing: Bonded-in technology for structural timber. Proc Inst Civil Eng Constr Mater 160(3):95–98. https://doi.org/10.1680/coma.2007.160.3.95

31. Bank LC (2006) Composites for construction: structural design with FRP materials. John Wiley & Sons

32. De Luca A, Nanni A (2011) FRP-reinforced concrete structures. In: Luigi N, Assunta B (eds) Wiley encyclopedia of composites, 2nd edn. John Wiley & Sons, Inc. https://doi.org/10.1002/9781118097298.weoc098

33. Serrano E (2001) Glued-in rods for timber structures - a 3d model and finite element parameter studies. Int J Adhes Adhes 21(2):115–127. https://doi.org/10.1016/S0143-7496(00)00043-9

34. Volkersen O (1938) Die Nietkraftverteilung in zugbeanspruchten Nietverbindungen mit konstanten Laschenquerschnitten. Luftfahrtforschung 15(1/2):41–47

35. Volkersen O (1953) Die Schubkraftverteilung in Leim-, Niet-und Bolzenverbindungen, Teil 1-3. Energie und Technik 5(3):68–71, 5(5):103–108, 5(7):150–154

36. Hassanieh A, Valipour HR, Bradford MA, Jockwer R (2018) Glued-in-rod timber joints: analytical model and finite element simulation. Mater Struct 51(3):61. https://doi.org/10.1617/s11527-018-1189-9

37. Gustafsson PJ (1987) Analysis of generalized Volkersen-joints in terms of non-linear fracture mechanics, Paper 20-18-2. In: Proceedings of the 20th conference of CIB-W18, Dublin, Ireland

38. Ottosen NS, Olsson KG (1988) Hardening/softening plastic analysis of adhesive joint. J Eng Mech 114(1):97–116. https://doi.org/10.1061/(ASCE)0733-9399(1988)114:1(97)

39. CEN. EN 1995-1-1 (2004) Eurocode 5: design of timber structures—part 1-1: general—common rules and rules for buildings. European Committee for Standardization CEN, Bruxelles, Belgium

40. Aicher S, Stapf G (2017) Eingeklebte Stahlstäbe – state-of-the-art – Einflussparameter, Versuchsergebnisse, Zulassungen, Klebstoffnormung, Bemessungs- und Ausführungsregeln. In: Proceedings of the 23rd Internationales Holzbau-Forum IHF 2017, Garmisch-Partenkirchen, Germany

41. DIN EN 1995-1-1/NA (2013) National annex—nationally determined parameters—Eurocode 5: design of timber structures—Part 1-1: general—common rules and rules for buildings, DIN Deutsche Institut für Normung e.V., Berlin, Germany
42. Bengtsson C, Kemmsies M, Johansson CJ (2000) Production control methods for glued-in rods for timber structures. Proceedings of the World conference on timber engineering, WCTE, Paper no. 7-4-1
43. Cimadevila JE, Chans DO, Gutiérrez EM, Rodríguez JV (2012) New anchoring system with adhesive bulbs for steel rod joints in wood. Constr Build Mater 30:583–589. https://doi.org/10.1016/j.conbuildmat.2011.12.052
44. Bengtsson C, Johansson C-J (2001) Glued-in rods—development of test methods for adhesives. In: Aicher S, Reinhardt H-W (eds) International RILEM symposium on joints in timber structures, RILEM Publications s.a.r.l., volume PRO22, Stuttgart, Germany. RILEM Publications
45. Clorius CO, Pedersen MBU, Damkilde L, Hoffmeyer P (1996) The strength of glued-in bolts after 9 years in situ loading. In: Proceedings of the international wood engineering conference, pp 496–503
46. Tannert T, Zhu H, Myslicki S, Walther F, Vallée T (2017) Tensile and fatigue investigations of timber joints with glued-in FRP rods. J Adhes 93(11):926–942. https://doi.org/10.1080/00218464.2016.1190653
47. Myslicki S, Bletz-Mühldorfer O, Diehl F, Lavarec C, Vallée T, Scholz R, Walther F (2019) Fatigue of glued-in rods in engineered hardwood products—Part I: experimental results. J Adhes 95(5–7):675–701. https://doi.org/10.1080/00218464.2018.1555477
48. Bengtsson C, Johansson C-J (2000) Test methods for glued in rods for timber structures. In: Proceedings of the CIB W18 Meeting 33, Delft, The Netherlands, Paper no. 33-7-8

Fiber-Reinforced Polymers as Reinforcement for Timber Structural Elements

Bo Kasal and Libo Yan

Abstract This chapter provides a comprehensive summary of recent developments in the use of fiber-reinforced polymers (FRP) as reinforcement for timber elements and joints. The constituents of FRP composites are first introduced. Next, the typical applications of FRP in the reinforcement of wood elements and joints are discussed. The long-term performance of adhesively bonded FRP-reinforced wood structures under different environmental effects (i.e. moisture, temperature, coupled moisture and temperature, and fire) are reviewed and discussed. Furthermore, the in situ quality control for the bonding of FRP to wood substrate and the assessment of FRP-reinforced wood structures are described. Finally, perspectives on the application of such systems are given.

1 Introduction

Fiber-reinforced polymers (FRP) have been used as reinforcement for wood structural elements to increase stiffness and/or strength and to control crack initiation and development.

Wood reinforcement by FRP is either applied *globally* or *locally*. *Global reinforcement* is usually in the form of a composite fabric located on the surface of a wood element. *Local reinforcement* is applied to increase the resistance of a wood material to locally concentrated stresses (or to control excessive strains); it can be applied either as textile reinforcement (surface) or as glued-in rods (as discussed in detail in Sect. 3.2) or pipes.

B. Kasal · L. Yan (✉)
Fraunhofer Institute for Wood Research, Wilhelm-Klauditz-Institut WKI, Braunschweig, Germany
e-mail: libo.yan@wki.fraunhofer.de

Department of Organic and Wood-based Construction Materials, TU Braunschweig, Braunschweig, Germany

B. Kasal
e-mail: bohumil.kasal@wki.fraunhofer.de

© RILEM 2021
J. Branco et al. (eds.), *Reinforcement of Timber Elements in Existing Structures*,
RILEM State-of-the-Art Reports 33,
https://doi.org/10.1007/978-3-030-67794-7_4

In order to select appropriate reinforcement, several conditions must be met. These include compatibility between the reinforcing and reinforced materials (e.g. under thermal and moisture effects) and ensuring that the adhesive layer is sufficiently thick and durable. In addition, time-dependent material properties of the FRP, polymeric matrix, adhesive and wood must be considered. The literature [e.g. 1–12] covers the conceptual function of wood reinforcement relatively well (i.e. large-scale tests of beams and columns, for example), but the fundamental questions related to ageing processes, long-term performance under various environmental conditions and mechanisms of deterioration are only barely investigated and addressed. Further, the interfacial behaviour of the composite cross-sections is poorly described.

This chapter provides a comprehensive summary of recent developments regarding FRP-reinforced wood elements and joints. The topics covered include the constituents of FRP composites, typical applications of FRP in the reinforcement of wood elements and joints, long-term performance of adhesively bonded FRP-reinforced wood structures under different environmental effects (i.e. moisture, temperature, coupled moisture and temperature, and fire), in situ quality control for bonding of FRP to wood substrate, and in situ assessment of FRP-reinforced wood structures. In addition, more perspectives on the application of adhesively bonded FRP-reinforced wood systems are also given.

2   FRP Materials

FRP is a composite material made up of reinforcing fibers (e.g. in the form of yarns and textile structures) and a polymeric matrix in which the fibers are embedded, with an interface between them. The fibers can be either man-made (e.g. glass, carbon, aramid) or natural (i.e. lignocellulosic, such as flax, hemp, coir, cotton and other plant-based fibers) [13]. A combination of man-made (endless fibers and/or yarn) and natural fibers (must be spun into yarn) is theoretically possible but rare [14]. Individual monofilament fibers are not used for wood reinforcement; rather, they are used in the form of rods, pipes, or fabric configurations. In FRP reinforcing rods, usually manufactured by the pultrusion process, the fibers are oriented parallel with each other and embedded in a polymeric matrix, mostly epoxy-based thermoset. For pipes and fabric, the system can have a more complex structure. While for a fabric stitch-bonding (i.e. non-woven fabric) and weaving are the main technologies used to fabricate a suitable textile, FRP reinforcing pipes are usually made by the knitting technology. The reinforcing layers of a fabric can have various architectures of the individual layers and offer almost an endless number of possibilities to control the mechanical behaviour of the fabric or a pipe [8]. By tailoring the architecture of a fiber fabric, one can control the mechanical parameters such as elongation, yield at failure, strength, and mechanical behaviour of the resulting system [15]. For instance, by inclining the fibers or yarn with respect to the principal loading direction, one can induce shear deformation (or stress) under normal loading. Ample literature dealing with fibers and their processing and production of textiles exist, e.g. [16].

In FRP, the fibers provide stiffness, strength and fracture toughness to the composite. The polymeric matrix holds the fibers together to form a shape and transfers the stresses to the fiber reinforcement by adhesion and/or friction [13]. In addition, the polymeric matrix acts as a barrier to protect the embedded fibers from outer environmental attack and corrosion. The polymeric matrix also has a great effect on the compressive, shear and transverse strengths of the FRP composite [13].

2.1 Fiber

In general, fiber materials can be classified into two categories: natural and man-made. Man-made fibers can be further divided into two sub-categories, i.e. organic and inorganic. Currently, man-made fibers dominate as reinforcing materials of FRP composites for various engineering applications. For civil infrastructure application, glass and carbon are the two most widely used man-made fibers. The primary advantages of using glass fibers in civil engineering are their relative low cost and high strength, as well as typical glass properties such as hardness, corrosion resistance and chemical inertness [17]. The drawbacks include the poor abrasion resistance of glass fibers, which is why they require the use of protective coatings during manufacturing, and the relative low modulus of elasticity of glass compared to the conventional steel rebar reinforcement. In comparison with glass fibers, carbon fibers are less widely used due to their much higher cost. Carbon fibers have extremely high strength and moderately high stiffness compared with other materials (e.g. steel), which make them ideal for lightweight structures. In civil infrastructure applications, where relatively large amounts of materials are required, glass is the more preferable because it is less expensive compared with carbon. However, for some applications where light weight and high strength are needed, e.g. in the case of seismic retrofitting of RC columns using FRP wrapping, carbon is more popular and cost-effective due to its superior strength and stiffness [17].

Based on their origin or botanical type, natural fibers can be classified into plant-based natural fibers (i.e. lignocellulosic), animal fibers (e.g. wool), and mineral fibers (e.g. basalt) [17]. Basalt is one of the natural mineral fibers that have been widely developed in recent years; its cost is equal to or even less than that of glass, but its strength is comparable to the strength of carbon. Basalt fibers have a better balance between strength and price when compared with glass and carbon [18]. In the past decade, the use of plant-based natural fibers as reinforcing materials in polymer composites for civil infrastructure applications has also gained popularity because of the requirement for sustainability in the built environment [19]. The main advantages of using plant-based natural fibers are their comparable high specific strength and stiffness to those of glass fibers, their little energy of production, their low CO_2 emission during production, and their low cost. Furthermore, the plant-based natural fibers are biodegradable with good thermal and acoustic insulating properties [18, 19]. Figure 1 gives the classification of fiber materials used in FRP composites [13, 18].

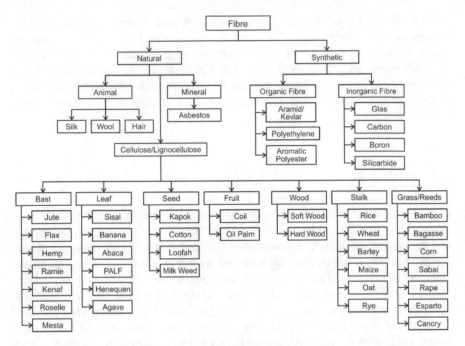

Fig. 1 Classification of fibers (reproduced with permission from [13])

Table 1 lists the tensile properties of several natural and man-made fibers obtained from single fiber tensile tests [13, 14, 18, 19]. Overall, the tensile strengths and moduli of man-made fibers and natural basalt fiber are much higher than those of plant-based natural fibers. However, the densities of plant-based natural fibers are lower than those of the man-made and basalt fibers. Among the man-made and natural basalt fibers, E-glass fibers have the lowest tensile strength and modulus. When considering fiber densities, it can be concluded that the specific strength and modulus of plant-based natural fibers such as flax will be comparable to those of E-glass fiber [19].

In a polymeric matrix, the fiber materials can be of different forms such as monofilament fibers, rovings, yarns, mat, and fabric textiles. Fibers are the most widely used as reinforcing materials in polymeric matrices for structural reinforcement in the form of textile. Normally, the tensile strength and elastic modulus of man-made fiber-fabric-reinforced polymer composites are close to the corresponding fiber properties obtained from single fiber tensile tests. For plant-based natural fiber-fabric-reinforced polymer composites, though, their tensile properties usually are much lower than the corresponding fiber tensile properties obtained from the single fiber tensile test. This can be attributed to the different tensile failure mechanisms of plant-based natural fabric- and man-made fabric-reinforced polymer composites in tension. In the latter, the mechanism of failure is similar to that of a single man-made fabric fiber in tension (i.e. fiber breakage), because man-made fabric fibers are continuous long

Table 1 Properties of natural and man-made fiber materials [13, 14, 18, 19]

Fiber type	Density (g/cm³)	Tensile strength (MPa)	Elastic modulus (GPa)	Elongation at failure (%)
Bamboo	0.6–1.1	140–800	11–32	2.5–3.7
Coir	1.15–1.46	95–230	2.8–6	15–51.4
Cotton	1.5–1.6	287–800	5.5–12.6	3–10
Flax	1.4–1.5	343–2000	27.6–103	1.2–3.3
Hemp	1.4–1.5	270–900	23.5–90	1–3.5
Jute	1.3–1.49	320–800	30	1–1.8
Kenaf	1.4	223–930	14.5–53	1.5–2.7
Ramie	1.0–1.55	400–1000	24.5–128	1.2–4.0
Sisal	1.33–1.5	363–700	9.0–38	2.0–7.0
Aramid	1.4	3000–3150	63–67	3.3–3.7
Carbon	1.4	4000–5000	200–240	1.4–1.8
E-glass	2.5	1000–3500	70–76	0.5–1.5
Basalt	2.7	3000–4840	79.3–93.1	2.0–3.5
Kevlar 29	1.44	2900	70	4.0
Kevlar 49	1.45	2900	135	2.8
Kevlar 129	1.45	3400	99	3.3

Note For E-glass, "E" standards for enhanced electrical properties. Kevlar® fiber and filament come in a variety of types, each with its own unique set of properties and performance characteristics for different protection needs. In general, Kevlar 29 is used in ballistic applications, ropes and cables, in protective apparel such as cut-resistant gloves, in life protection applications such as helmets, vehicular armoring, and plates, and as rubber reinforcement in tires and automotive hoses. Kevlar 49 is used primarily in fiber optic cables, textile processing, plastic reinforcement, ropes, cables, and composites for marine sporting goods and aerospace applications. Kevlar 129 is used in motorcycle racing gears, life protection accessories, ropes and cables, and high-pressure hoses used in the oil and gas industry [78]

fibers (endless fibers) in synthetic textiles. For plant-based natural fabric composites, the tensile failure is a combination of fiber slippage and fiber breakage, which is different from the fiber breakage failure of single plant-based natural fibers in tension. When spinning the short plant-based natural fibers into yarns, a number of fiber filaments are twisted into a continuous strand, producing radial forces that cause the movement of some of these filaments relative to others, which leads to a closer packing of all the filaments within any given cross-section [20]. The tensile strength of the fiber bundle cannot be as high as that of the fiber strand yarn because, close to failure, some fibers break and others slip. The tensile failure of plant-based natural fiber yarn is dependent on the yarn structure, i.e. the configuration, alignment and packing of constituent fibers in the yarn cross-section. For plant-based natural fiber fabric with loose packing of fibers in the yarns, the yarn failure mechanism is slippage-dominated, thus the load-bearing capacity of the slipped fiber is drastically reduced and the final yarn strength is poor, resulting in poor strength of such fabric-reinforced polymer composites [20].

2.2 Polymeric Matrix

Polymeric matrices are typically classified into two categories, depending on how they response to heat: thermoplastics and thermosets. Thermoset matrices undergo permanent reactions when heated, while thermoplastic matrices undergo only temporary physical changes. Thermosets are formed by irreversible chemical reactions between polymeric chains (i.e. cross-linking). Thermoplastics consist of long polymeric chains that can be heated, deformed, and cooled with no occurrence of chemical reaction [17].

The most commonly used thermoplastic matrices are polypropylene, polyethylene and polystyrene. Table 2 lists the properties of these typical thermoplastics [19]. When heated, thermoplastics do not cure or set. They soften and become more fluid when temperature goes up. Therefore, the processing of a thermoplastic involves softening and melting when heat is applied. When the thermoplastic cools, it hardens into the shape of its mould. No chemical reaction or cross-linking occurs and the changes in this kind of polymer are entirely physical. Typically, thermoplastics are more ductile than thermosets. The failure strains of thermoplastics can be much larger than 100%, whereas the failure strains of thermosets are normally less than 10% [17]. In civil engineering, thermoplastics are rarely used due to their poor chemical resistance, high shrinkage (which can be as high as 20%), and poor durability. Additionally, thermoplastics normally have lower stiffness and strength than thermosets. However, thermoplastics have some advantages over thermosets such as low melting point, ease of processing, recyclability and good fracture toughness (i.e. ductility) [19].

Table 2 Properties of typical thermoplastic polymers [19]

Property	PP	LDPE	HDPE	PS
Density (g/cm^3)	0.90–0.92	0.91–0.93	0.94–0.96	1.04–10.6
Water absorption (24 h@20 °C)	0.01–0.02	<0.015	0.01–0.2	0.03–0.10
T_g ($^\circ$C)	−10 to −23$'$	−125	−133 to −100$'$	N/A
Tensile strength (MPa)	26–41.4	40–78	14.5–38	25–69
Elastic modulus (GPa)	0.95–1.77	0.055–0.38	0.4–1.5	4–5
Elongation (%)	15–700	90–800	2.0–130	1–2.5
Izod impact strength (J/m)	21.4–267	>854	26.7–1068	1.1

Note *PP* polypropylene, *LDPE* low-density polyethylene, *HDPE* high-density polyethylene, *PS* polystyrene, T_g glass transition temperature

The common thermoset resins are polyester, vinyl ester, and epoxy, as listed in Table 3 [19]. Thermosets are the most widely used polymeric matrices in civil engineering because they outperform thermoplastics in terms of mechanical properties, chemical resistance, thermal stability and overall durability. In addition, thermoset matrices tend to have better interfacial bonds between the polymeric chains. Furthermore, thermosets allow for more flexibility in their structural fiber configurations and for processing at room temperature [19]. Epoxy resin is currently the best generic adhesive type or polymeric matrix used for wood and FRP adhesive bonding.

A great variety of formulated commercial epoxy adhesives is available, offering a wide range of properties, such as adhesion to wood, viscosity, curing time and shrinkage, tensile strength and modulus. The application of different types of epoxy adhesives does not require high pressure and temperature when curing, and epoxy adhesives are reasonably tolerant to glue-line thickness variations [2]. In addition, they also show strong adhesion of dissimilar materials (e.g. wood and concrete or wood and metal), low or even no shrinkage during curing, good dimensional stability after hardening, good mechanical resistance, and good chemical and water resistance [2]. Unlike traditional generic adhesive types, the epoxy adhesives can be mixed and processed to cure under a wide variety of ambient conditions. This is an essential requirement for in situ application. Due to their ability to adhere strongly to other materials besides wood, epoxy adhesives can also be used in conjunction with FRP plates and FRP rods for the reinforcement of wood structures [2].

The mechanical properties of all polymeric materials are highly affected by the processing temperature [17]. When a thermoset (e.g. epoxy) is exposed to elevated temperatures close to but still less than the glass transition temperature of its matrix, T_g, its mechanical properties (strength and stiffness) and long-term durability decrease significantly. T_g is defined as the temperature at which a material transitions from a hard and brittle state to a soft and rubberlike state [17]. If a thermoset is exposed to a temperature close to the T_g of its matrix, there will be a dramatic reduction in its stiffness. The mechanical properties of thermosets are

Table 3 Properties of typical thermoset polymers [19]

Property	Epoxy	Polyester	Vinyl ester
Density (g/cm³)	1.1–1.4	1.2–1.5	1.2–1.4
Water absorption (24 h@20 °C)	0.1–0.4	0.1–0.3	0.1
T_g (°C)	65–105	60–85	–
Tensile strength (MPa)	35–100	40–90	69–83
Elastic modulus (GPa)	3–6	2–4.5	3.1–3.8
Elongation (%)	1–6	2	4–7
Izod impact strength (J/m)	0.3	0.15–3.2	2.5

also time-dependent, which causes viscoelastic stress-strain response that makes the matrices susceptible to creep.

3 FRP Reinforcement in Wood Structures

FRP reinforcement for wood structures can be divided into two types, based on the configuration: (i) pultruded FRP products (rods and plates) and (ii) fabrics. Usually, pultruded FRP rods and plates are utilized as internal reinforcement and bonded into slots or grooves in a wood component [1]. Fabrics are normally used as external reinforcement. For both types of use, the implementation of the FRP reinforcement will be applied by the use of adhesive bonding.

Adhesive bonding is regarded as the most efficient technique for the transfer of stress between FRP reinforcement and wood substrates, as it prevents stress concentrations that are usually associated with mechanical fasteners [2]. In engineering application, a widespread utilization of adhesive bonding can be attributed to its inherent advantages, namely: (1) an adhesive joint can distribute the applied load over the entire bonded area with a more uniform distribution of stresses; (2) it causes little or no damage to the adherends; (3) it adds very little weight to the wood structure and has a superior fatigue resistance compared to other joining methods; (4) it is suitable for joining dissimilar materials; and (5) it can reduce manufacturing costs [2].

To use fabrics as reinforcement, a wet lay-up process is normally employed to apply an adhesive-impregnated fabric onto the surface of a wood component by means of a primer epoxy adhesive layer. For pultruded FRP plates, a primer adhesive, such as epoxy, is normally applied to a wood surface and then the pultruded FRP plates are glued to the wood component with the appropriate pressure. FRP rods are bonded and embedded into pre-drilled slots, holes or grooves with adhesives. Glued-in FRP rods are effective for achieving stiff, high-capacity connections in wood engineering. They can be used for in situ repair and strengthening, as well as for new wood construction. They can also be used for wood column foundations, for moment-resisting connections in beams and frame joints and corners, as shear connectors, and for strengthening structural components extensively loaded perpendicularly to the grain and in shear [7]. In the following subsections, typical examples of the application of different FRP reinforcement in wood structures are presented. The possible reinforcement techniques of FRP in wood are shown in Fig. 2. FRP reinforcement can be either surface applied or glued-in as rods or pipes.

3.1 FRP Reinforcement Along the Wood Fibers

The reinforcement of wood along fibers is facilitated by the adhesion of FRP to the surface of the reinforced wood element and/or embedment of FRP rods in pre-drilled

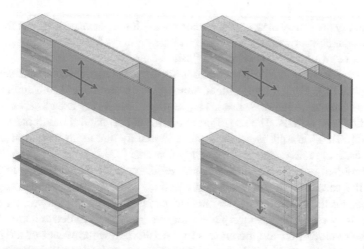

Fig. 2 Schematic diagrams of the reinforcement of wood with FRP

holes or grooves [4, 6]. The goal here is to increase the load-carrying capacity of the reinforced member by attaching a layer of FRP to the side with the expected tensile stress. By using a two-dimensional fabric, the reinforcement will result in a two-dimensional action (i.e. also reinforcing wood across fibers) but the dominant action is along fibers.

Unlike concrete reinforcement, where the concrete has negligible tensile strength compared to its compressive strength, the reinforcement of a wood member parallel to the grain does not assume cracked conditions and utilizes the entire cross-section. The observed tensile strains when the first cracks start to develop are below 1% [6], which means that the stiffness of the FRP reinforcement must be relatively high to transfer the released internal forces. Stress-strain tests of GFRP rods showed stresses between 250 and 350 MPa at about 1% of the strain. Tests performed in [7] resulted in stress of about 80 MPa at 0.2% of the strain, which, when extrapolated, would give about 400 MPa of stress at 1% of the strain. A relatively thick reinforcement layer is needed to achieve measurable effects on the stiffness.

Numerous sources have documented the improved global performance of FRP-reinforced beams (i.e. reinforcement usually at the tensile side of a simply supported beam) [9–12, 21]. The research focused, however, on the global performance, such as deformations or moment capacities (stiffness and strength) and, often, uses of transformed cross-sections, to evaluate the effectiveness of the reinforcement. Some authors did not find the reinforcement to be particularly effective, reporting little percentage increase (1.5–4%) in strength and stiffness due to FRP reinforcement [22]. The conclusions, however, depended on the nature of the reinforcement itself and could not be generalized.

FRP reinforcement can be applied to newly constructed beams or in the repair of damaged beams [23–27]. Most FRP reinforcement and repair techniques were conceptually proven by means of full-scale tests (global performance), but the underlying mechanisms governing, for example, long-term performance and/or behaviour in fire were only scarcely investigated [28].

FRP of relatively large cross-sections used as wood reinforcement were also investigated [29]. Such reinforcement, although not common, can find application in seismic retrofitting where low masses are required. Most FRP use glass, carbon, and Kevlar fibers or textiles. The use of other fibers such as natural basalt has also been reported [30–32]. Basalt fibers show promise mainly due to their good mechanical properties, inorganic nature, and relatively low cost [33].

A combination of shear and flexural reinforcement was reported in [34] to reduce the variability of the product properties, which is a significant component in increasing the reliability of the reinforced member. No reliability analysis was performed by the authors. In [35], the safety factors of reinforced and unreinforced beams were compared and increments in the 10th percentile values of the bending strengths were reported. This, however, cannot be generalised to all reinforcing strategies and such results are always specific to a particular design and configuration, including the boundary and loading conditions. In [36], the authors claim to have performed reliability analysis of reinforced laminated beams using first-order reliability method. Reliability analysis is only rarely performed but is, in the authors' opinion, needed to prove the plausibility of individual reinforcing strategies. FRP reinforcement tends to decrease the variability of solid wood beams and increase their capacity [37].

The effect of reinforcement on the position of the neutral axis as a function of the reinforcement ratio (defined as the area of reinforcement divided by the area of the beam) is shown in Figs. 3 and 4. The figures were generated for a prismatic beam with FRP reinforcement at the bottom face and full compatibility of the strain assumed. It is clear that to achieve a significant shift in the neutral axis towards reinforcement, a high reinforcement ratio is needed. For example, as illustrated in Fig. 4, when the ratio of the modulus of FRP reinforcement to that of wood is 50, a reduction in the ratio of the depth of the neutral axis to the depth of the beam results in a significant increase in the reinforcement ratio from 0 to 10%. Modelling of reinforced beams ranges from simple transformation of cross-sections to numerical techniques [38].

Limited fire tests were performed; a significant drop in the design strength of FRP when subjected to fire conditions was observed [39]. Experiments conducted in [40] led to the conclusion that "FRP reinforced glulam could be designed for fire rating in a similar manner as conventional non-reinforced glulam". The fire performance of FRP-reinforced wood structures should be a focus of future research.

3.2 FRP Reinforcement Across the Wood Fibers

The reinforcement of wood across fibers is facilitated by the adhesion of FRP to the surface of the reinforced element, glued-in rods or pipes running across the fibers.

Fig. 3 Shift in the neutral axis as a function of the reinforcement ratio and relative E-modulus of reinforcement expressed as a multiple of the modulus of elasticity of wood in tension parallel to the fibers

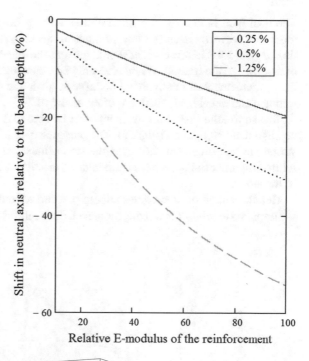

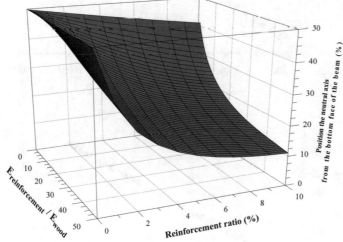

Fig. 4 Effect of reinforcement on the position of the neutral axis of a prismatic beam

The goal here is to control brittle failure resulting from crack initiation, development and growth. Given the low strength of wood in tension perpendicular to the fibers, glass FRP reinforcement should prevent the development and/or propagation of cracks. Tensile (radial) stresses can only be transferred through the interfaces and the reinforcement of relatively high stiffness, given the strains at which the crack is initiated and developed which are often below 0.2% [7, 8], as illustrated in Fig. 5a. At that strain, the FRP reinforcement must be able to develop sufficient stress with no significant elongation (Fig. 5b). The presence of crack will significantly increase stresses in reinforcement. However, the strains must remain small, meaning that the reinforcing material must have a modulus of elasticity significantly higher than that of the wood.

Reinforcement of wood across fibers (i.e. further radial reinforcement) has also been applied to joints transferring moments. Such reinforcement can either be applied

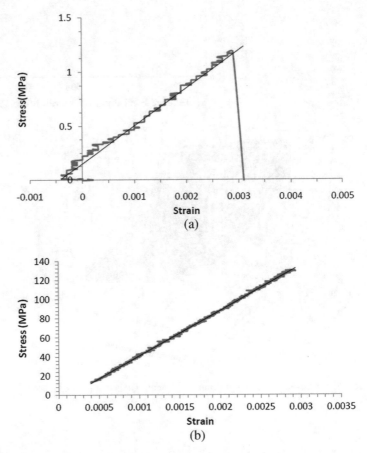

Fig. 5 Stress-strain diagram for spruce wood loaded in perpendicular tension. **a** Crack develops at a strain of about 0.3%; **b** Stress-strain diagram for a glass fiber reinforcing rod (reproduced with permission from [7])

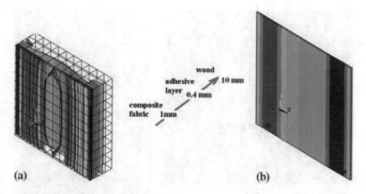

Fig. 6 Schematic of the reinforcement of wood with FRP—radial reinforcement to control cracks. **a** Tensile stress contours on the reinforcing fabric surface; and **b** principal stress in the epoxy adhesive layer—an example of the isoparametric elements and modelling of the static load applied to wood layer (reproduced with permission from [39])

to the surface of connected elements or, again, as rods inserted in pre-drilled holes [37, 39, 41]. Surface reinforcement to control cracks is functional in laboratory conditions and for relatively narrow beams or columns. Surface reinforcement should not, effectively, be relatively far from the surface; even in the adhesive layer strain decay can be expected [37]. Parameters, such as the effective depth of surface reinforcement and the spacing of internal reinforcing elements (such as glued-in rods) to control internal crack development, are yet to be determined, both theoretically and experimentally.

Figure 6 shows the results of the analysis of the interface stress when glass fiber (GF) fabric is used to reinforce wood loaded by tension perpendicular to the grain direction. Figure 7 shows the reinforcement of glue-laminated columns with GF rods to mitigate crack development due to moments transferred.

3.3 Repair of Damaged Structural Components

The reinforcement of wood along fibers is facilitated by the adhesion of FRP to the surface of the reinforced element. The goal here is to increase the loading-capacity of the reinforced member by attaching a layer of FRP to the side with the expected tensile stress. Using FRP for repair is relatively common and a number of sources cover this topic, e.g. [1, 23, 25, 26]. An example of repair of a damaged wood bridge with CFRP reinforcement is shown in Fig. 8 [42]. Again, evidence of the functionality of the reinforcement is mostly at the global, proof-of-concept level, and questions related to the compatibility of reinforcing and reinforced materials, behaviour of the adhesive interface under thermal and moisture fluctuations, as well as long-term performance and fire resistance are yet to be investigated.

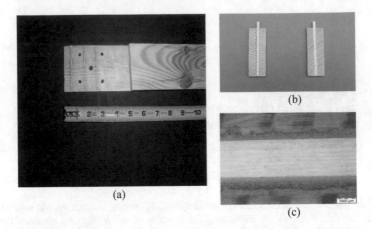

Fig. 7 Radial reinforcement of wood moment connection with GF glued-in rods. **a** reinforcement around the dowels, **b** reinforcing rods running across the grain, and **c** microscopic image of the reinforcement (epoxy glue-line is clearly visible)

4 Long-Term Performance of FRP-Reinforced Wood Structures

Although FRP reinforcement is effective in the enhancement of the capacity and stiffness of wood structural members, the long-term performance has still not been demonstrated since the function of the reinforcement is mostly proven via short-term tests and data on the long-term performance are insufficient. The long-term performance is often investigated through classical creep tests, which cannot simulate real in situ conditions. In reality, the fluctuations in load, temperature and relative humidity are random phenomena, while most of the experiments are deterministic in nature, frequently carried out under constant relative humidity and temperature. Using polymers in FRP (wood itself is frequently defined as a polymer) offers the opportunity to use time-temperature superposition as a method to define viscoelastic properties of individual components, although a superposition of results for multiple materials (even if all components are polymers) is faced with theoretical challenges.

In the literature, a large number of factors influencing the long-term durability of adhesively bonded FRP-reinforced wood structures have been identified. These factors can be mainly divided into three categories: environment, materials and stresses [2]. In this section, the environmental effects, including moisture, temperature, coupled moisture and temperature, and fire, on FRP-reinforced wood structures are reviewed and discussed.

Fig. 8 Sins bridge in
Switzerland: **a** Installation of
CFRP on the lateral wood
beam in 1992; **b** long-term
monitoring of the
FRP-repaired wood beam in
2006 (reproduced with
permission from [42])

4.1 Effect of Moisture

In general, when subjected to water, adhesively bonded FRP-wood systems can lose
their bond strengths and, consequently, their load-carrying capacities over time. In
extreme conditions, the bonded FRP-reinforced wood may lose its function due to
severe degradation at the FRP and wood interface such as delamination [43]. The
transportation of water to the FRP-wood interface or in the FRP-wood system is
facilitated by four different mechanisms: (1) diffusion inside the polymeric adhesive,
(2) diffusion inside the permeable wood, (3) imperfections within the polymeric
matrix (i.e. formation of microspaces, pores and cracks), and (4) capillarity along
the reinforcing fiber/adhesive and wood/adhesive interfaces [13].

When an FRP and the FRP/wood interface are exposed to moisture, water
surrounding the reinforcing fiber/adhesive and wood/adhesive interfaces penetrates

and attaches onto the hydroxyl groups of the wood fibers and the reinforcing fibers (e.g. glass or plant-based natural fibers) of the FRP. This process can create intermolecular polar bonding with the wood and even with the man-made or plant-based reinforcing fibers (i.e. in FRP). This process can reduce the interfacial bond between the reinforcing fibers and the polymeric adhesive and between the polymeric adhesive and the wood components. The differential swelling/shrinkage causes strains within the wood-adhesive-FRP system, which in turn may cause development of cracks. These cracks further promote capillarity and transport of water via microcracks. Afterwards, water-soluble substances start leaching from the wood cellulosic fibers and/or reinforcing fibers, and this eventually results in debonding between the wood and the adhesive matrix and between the reinforcing fiber and the adhesive matrix. Debonding between the fiber and polymeric adhesive is initiated by the development of osmotic pressure pockets at the surface of the fibers due to the leaching of the water-soluble substances from the surface of the wood cellulosic fiber [13, 44]. Figure 9 shows the schematic view of the effect of water on the interfacial bond of cellulosic fiber (i.e. wood) and polymeric adhesive (i.e. epoxy adhesive).

The degradation of the FRP and wood interfacial bond due to penetration of water into their interface might be attributed to several reversible and irreversible mechanisms. At the initial stage of penetration, the effect of water on the FRP and wood interface might be reversible, which is true if the surface of the wood substrate is well prepared and treated and hydrolytically stable adhesive is applied [45]. When

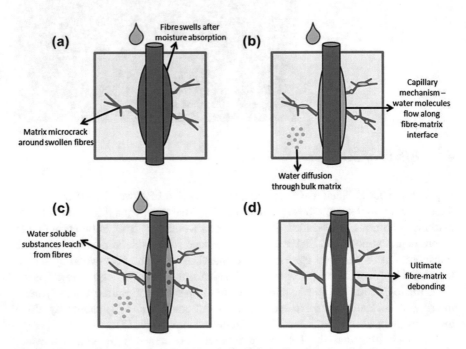

Fig. 9 Effect of water on the interfacial bond of cellulosic fiber (e.g. wood) and a polymer matrix (e.g. adhesive) (reproduced with permission from [13])

the FRP and wood become dry, the strength lost at their interfacial bond may be partially recovered. The irreversible processes may occur over a period of time at the critical moisture concentration (which is defined as the average moisture throughout a solid material being dried, its value being related to drying rate, thickness of material, and the factors that influence the movement of moisture within the solid [45]). Above this critical concentration, the weakening at the FRP/wood interface (or joint) may become irreversible. It is believed that the critical water concentration is highly influenced by the materials that form the joint (i.e. FRP, wood and adhesive for a bonded FRP-reinforced wood system), temperature and the applied stress conditions in the joint. With time, an accumulation of the damage induced by the irreversible mechanisms at the FRP and wood interfacial bond will cause a complete failure of the interface.

Davalos et al. [46] evaluated the in-service durability of phenolic-bonded FRP-wood interfaces produced by face bonding and of epoxy-bonded FRP-wood interfaces produced by filament winding. This study showed that, under ambient moisture conditions, both adhesively bonded FRP-wood interfaces degraded and so did their long-term durability. The existence of moisture or water can reduce the bond strength of an FRP-wood interface and diminish the strengthening effectiveness of adhesively bonded FRP-reinforced wood, as confirmed by macro-scale [47] and micro-scale [48] characterization. A study combining macro-scale experimental tests and nano-scale molecular dynamics (MD) simulation [49] showed that the adhesion energy measured by MD for epoxy/wood interface exposed to a wet environmental condition dropped to one-third of that measured for the epoxy/wood interface under a dry condition. The results indicated that the bond strength between the wood and the epoxy decreased with higher moisture, and the reduction was even greater for a concrete and epoxy interface. The MD simulation also demonstrated that the presence of water molecules hindered the attraction between the epoxy adhesive and the wood substrate in the FRP-bonded wood beams. Furthermore, the load-carrying capacity of epoxy-bonded carbon FRP-wood samples decreased significantly with an increase in the exposure duration of the samples to a humid environmental condition. Although some studies, e.g. [46, 47, 50], have investigated the effects of different moisture conditions on FRP and wood interfacial bonds and the overall structural performance of bonded wood structures, studies on the effect of moisture on long-term bond performance and structural performance are still needed to better understand bonded FRP-wood hybrid systems during their service life. In addition, the stochastic natures of loading and environmental factors need to be experimentally and theoretically addressed, including the interaction between individual variables.

4.2 Effect of Temperature

In addition to moisture, temperature has a significant effect on the long-term performance of an adhesively bonded FRP-reinforced wood structure because it can result in the deterioration of the wood substrate itself and the adhesive at the FRP and wood interface. The effect of temperature on the bond strength of bonded FRP-reinforced wood can be classified into two categories: (1) the effect of temperature variations due to natural service environment; and (2) due to a fire (as discussed in Sect. 4). In the first category, the expected reasonable temperature variations are in the range from −18 to 65 °C. As reported in [43], temperature of a wood material surface in the range between 80 and 100 °C may be considered safe for prolonged heat exposure. When the temperature of the wood surface exceeds 100 °C, permanent strength degradation may result. At such a high temperature level, the FRP and wood interfacial bond strength can experience degradation if the polymeric adhesive used has a low glass transition temperature, as is the case for polyurethane (PUR) and epoxy adhesives [50]. Richter and Steiger [50] found that a crosslinked polymer of PUR and epoxy adhesives displayed significant viscoelastic responses over temperature ranges relevant to practical applications. Temperature-dependent creep was a risk factor that needed to be considered for structural application of these adhesives.

Polymeric adhesives can soften and become viscous when the temperature is in the range between 60 and 150 °C [51, 52]. They tend to lose their properties, e.g. stiffness, in certain temperature ranges, such as from 80 to 100 °C [53]. This is because the glass transition temperatures of most widely used polymeric adhesives (e.g. epoxy) are below or around 100 °C. Therefore, these adhesives, such as epoxy, are very sensitive to fire. However, it should be pointed out that some commercial polymeric adhesives, such as phenol-resorcinol and melamine-based adhesives (e.g. melamine-resorcinol formaldehyde and melamine-urea formaldehyde), which are used for wood, exhibit relatively good resistance at elevated temperatures [54].

Other studies [2, 45, 55] have pointed out that at high or low temperatures, the adhesives used in adhesively bonded FRP-reinforced wood may experience severe internal stresses that develop because of the different coefficients of thermal expansion of FRP and wood. In addition, adhesives also tend to get soft at elevated temperatures and brittle at low temperatures. Their long-term exposure to elevated temperatures can also lead to their oxidation or pyrolysis. Lartigau et al. [56] investigated experimentally and numerically the effect of temperature on the structural performance of glued-in rods in wood. The specimens with adhesively bonded glued-in rods were heated in an oven until the glue line reached a target temperature (i.e. 30, 40, 60 or 80 °C), and then a pull-out test was conducted. The results showed that there was a significant decrease (up to 45%) in the adhesive bonding strength when the adhesive was exposed to a temperature above 60 °C, where the glass transition temperature of the epoxy adhesive used was 58 °C. This implies that the inherent properties of the epoxy adhesive had a significant effect on the mechanical performance and the durability of the glued-in rods for wood reinforcement. A recent study [57] reported that low temperatures (i.e. 0–20 °C) had an insignificant effect on the long-term durability of the interfacial bond between an FRP and wood substrate. However, at temperatures between 20–50 °C, the adhesive displayed an enhanced viscoelastic response, which increased the risk of creep of the adhesively bonded FRP-wood hybrid system. When exposed to temperatures not less than 50 °C, the interfacial bond experienced partial degradation, particularly when the temperature of the FRP/wood interface was close to the glass transition temperature of the epoxy adhesive. Figure 10 shows the different failure modes of adhesively bonded carbon

Fig. 10 The different failure modes of adhesively bonded carbon FRP-wood samples after they were subjected to temperature levels of **a** 5 °C, **b** 20 °C, and **c** 50 °C for 12 weeks (reproduced with permission from [57])

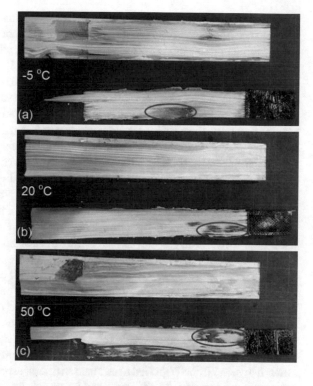

FRP-wood samples after they were subjected to increasing temperature levels (5, 20 and 50 °C) for 12 weeks [57]. The interfacial deterioration is indicated by the red circles. Wood delamination was the major fracture pattern among the three cases. The interfacial deterioration at 50 °C was more serious compared to those at 5 and 20 °C. As shown in Fig. 10, the separation area between FRP and wood at the highest temperature (i.e. 50 °C) is the largest among all the tested specimens. Therefore, the effect of service temperature on the bond performance of adhesively bonded FRP-wood systems should always be considered in the structural design. When applying adhesively bonded FRP-reinforced wood systems in tropical or hot climate regions where the sunlight exposure is common, extra attention should be paid to the temperature effect.

4.3   Effect of Combined Moisture and Temperature

The coupled effect of moisture and temperature is known to significantly accelerate creep. Although this phenomenon has been observed experimentally, theoretical explanations are still speculative in nature. Nardon et al. [58] investigated the influence of moisture and thermal cycles on the bond performance of epoxy adhesively bonded FRP (i.e. glass, carbon and aramid) strengthened wood (i.e. spruce) elements.

The results showed that the different fibers were sensitive to specific environmental conditions of temperature and humidity. The moisture content of wood did not significantly influence the success of the reinforcement for all fibers. CFRP and Aramid FRP (AFRP) are very sensitive to cyclical variations in temperature and humidity. High temperature has a negative effect on all FRP materials. In particular, it was noted that temperatures higher than 80 °C resulted in a substantial drop in the tensile performance of FRPs, particularly carbon and glass FRPs, each of which showed a reduction of more than 30% in strength when exposed to a temperature of 120 °C. The AFRP composite did not show any evident drop in tensile strength at high temperatures, but generally showed a global loss of strength (up to 25%) compared to the other two fibers. The prolonged presence of high humidity resulted in the decrease in strength. In particular, AFRP showed a drop in tensile strength of 35% at a moisture content of about 40%. The data obtained in this study showed that the more aggressive condition is prolonged high humidity, and that there is a gradual deterioration with time. Korta et al. [59] investigated the effect of humidity-temperature cycling on structural multi-material adhesive joints, i.e. CFRP-epoxy-wood. They concluded that even moderately harsh humidity-temperature loads caused debonding even if no external forces were applied. A series of finite element analyses were also performed to simulate the exposure of the samples to the chosen environmental conditions. The temperature expansion coefficient was identified as the crucial factor on which the performance of the joints made from the dissimilar materials depended. Stoeckel et al. [60] found that moisture typically caused the softening of adhesives and temperatures between 20 and 70 °C led to reduced stiffness, especially for polyvinyl acetate, polyurethanes and epoxy adhesives used for wood. The combination of thermal and humidity cycles caused significant reduction in CFRP and wood interfacial bond strength, as reported in [48]. Another study pointed out that the coupled effect of temperature and freeze/thaw cycling conditions could even lead to debonding of FRP laminates [61]. Thus, the long-term durability of adhesively bonded FRP-strengthened wood structures under the combined effect of moisture and temperature variations should be researched in future studies. Most of the cited studies treated the wood-adhesive-reinforcement system as a black box and the explanation for the deterioration at the molecular and/or atomic levels, especially with regard to the environmental effects, remains a challenge.

4.4   Effect of Exposure to Fire

Although there have been some studies on the fire performance of wood structures (e.g. [62, 63]), there is still very limited information available in the literature with regard to the fire performance of adhesively bonded FRP-reinforced wood systems, e.g. [62, 63].

Without protection from heat, a polymeric matrix may ignite, emit smoke, and facilitate the spread of fire. When exposed to fire, FRP materials may undergo charring, melting, delamination, cracking and deformation [64]. A fire test of GFRP-confined concrete columns [65] showed that FRP composites, when exposed to temperatures above 400 °C, were susceptible to the combustion of their polymer matrices and even evaporation. Igniting a polymer matrix at a high temperature could lead to the release of smoke, heat, and toxic fumes. The fiber materials could decompose and the FRPs could lose their confinement effects. As a result, concrete members reinforced by FRPs may not be able to perform their load-bearing functions. Compared with concrete, wood is considered a flammable material. Thus, the post-fire performance of FRP-strengthened wood structure might be even worse compared with that of FRP-strengthened concrete structures. When exposed to fire, the decomposition of the adhesives and/or the polymer matrix of the FRP may cause drastic degradation in the load-carrying capacity of the FRP-strengthened wood structures. For fire protection, it is believed that the encasing of FRP materials with wood elements (e.g. cladding) of sufficient thickness can provide the necessary fire resistance for the strengthened wood members [64]. This kind of fire protection method is quite suitable for near-surface mounted FRP systems for wood structure strengthening, which also provides necessary aesthetic requirements, as illustrated in Fig. 11. It has been proven effective for improving the fire performance of FRP-strengthened glulam (i.e. glued laminated wood) structure. Williamson and Yeh [66]

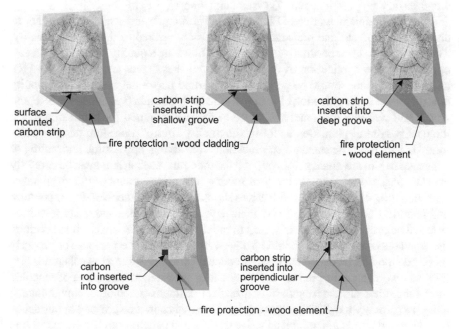

Fig. 11 Fire protection of FRP elements (reproduced with permission from [64])

investigated the fire performance of FRP-reinforced glulam, where the FRP laminates were embedded into the glulam body. The test results demonstrated that the glulam beam, with an FRP applied to the bottom face of the beam and directly exposed to fire, (no bumper lamination) could be designed to achieve a one-hour fire rating when evaluated in accordance with the ASTM E119 fire test protocol. There were no discernible differences in overall fire and structural performance between the two different FRP reinforcement layers (i.e. E-glass/epoxy and E-glass/urethane composites).

5 In Situ Quality Control and Assessment of FRP-Reinforced Wood Structures

5.1 In Situ Quality Control for FRP and Wood Bonding

A successful implementation of FRP reinforcement in wood structures is highly dependent on how well the system is known and good quality control at each implementation step. The performance of an adhesively bonded FRP-wood system can be controlled by considering various factors: (1) wood property, (2) adhesive property, (3) FRP property, (4) application of the adhesive bonding, (5) wood geometry, (6) wood surface preparation, and (7) system under service conditions.

For in situ implementation of FRP and wood bonding by adhesive, adhesive application and wood surface preparation are extremely critical to the bonding quality. These processes must be strictly implemented following a quality control plan developed before the application of the adhesive. All these steps involved in the FRP and wood bonding should be performed by skilled personnel based on the quality control plan, i.e. the selection of the most suitable materials (i.e. adhesive, FRP and wood) and working tools and adherence to the well-planned procedures based on the quality control plan. Details for in situ quality control procedure design can be found in [53, 67]. The standard specifications for mixing, application and testing of adhesives in situ are given in [68, 69]. All the materials used in situ must be carefully checked (e.g. expiry date, accuracy of mixing of the component of the given adhesive, etc.), properly stored (e.g. adhesives should be stored under certain temperature and humidity conditions, etc.), and their details recorded (i.e. the density, dimension and quantity of the materials, etc.) to satisfy the specifications. Before bonding, the wood substrate surface should also be well prepared. The reasons for properly prepared wood surface include: (1) to produce a close fit between the adherends (i.e. FRP and wood); (2) to produce a freshly cut or prepared surface, free of machine marks and other surface irregularities, free from extractives and other contaminants; and (3) to produce a mechanically sound surface that cannot be crushed or burnished (crushing and burnishing inhibit adhesive wetting and penetration) [2]. Normally, the preparation of wood surface involves some form of machining, either to create flat surfaces or specific mating shapes. In both situations, tight control of the fit between the mating surfaces is necessary to produce a bond line with uniform thickness. The quality of the surface varies with the type of the machining process as well as with how carefully the process is controlled [70].

In addition, the age of the wood surface prepared can also control the bonding quality. All wood surfaces immediately experience an inactivation process after preparation. The severity and rate of this inactivation is determined by the wood species, wood moisture content, temperature level and time of exposure to the temperature. An inactivated wood surface bonds weakly with adhesives because the inactivation process reduces the ability of an adhesive to properly wet, flow, penetrate and, in some cases, can impact cure, e.g. for phenolic and urea adhesives [71]. External contamination and self-contamination can cause chemical inactivation in wood surfaces. External contamination is caused by air-borne chemical contaminants, oxidation and pyrolysis of wood bonding sites from excessive drying or over-exposure to high temperatures, and impregnation with preservatives, fire retardants and other chemicals. Self-contamination results from a natural surface inactivation process where the hydrophobic wood extractives might migrate with time to the wood surface and can there undergo chemical reactions [72, 73]. Figure 12 shows multi-scales in adhesive bonds of solid wood. As can be seen, the bonding of wood with adhesives has a very complex structure at the microscopic level, even without the presence of FRP materials which also have complex microstructures. This makes the bonding of FRP and wood a very complex process.

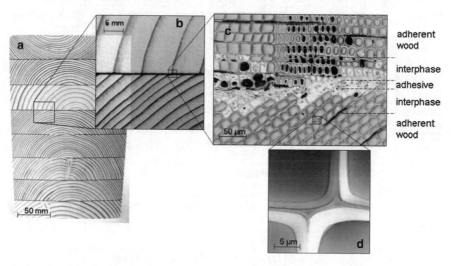

Fig. 12 Scales in adhesive bonds of solid wood: macroscopic scale (**a** and **b**); microscopic scale with indicated bond regions (**c**); atomic force microscopy image of wood cell walls (**d**) [60]

5.2 In Situ Assessment of Bonded FRP-Wood Structures

Under service conditions, it may be necessary to perform in situ assessment to monitor the bond performance of adhesively bonded FRP-wood structures. The current assessment techniques for bonded FRP-wood elements can be divided into three categories: non-destructive (NDT), semi-destructive (SDT) and destructive [74]. The NDT techniques are usually very efficient for rapid screening to identify potential locations of defects in wood structures using the correlations between NDT and destructive parameter. The effectiveness of various in situ NDT and SDT assessment techniques for FRP-reinforced wood structures is listed in Table 4 [75]. The theoretical basis, typical equipment set up, and basic capabilities and limitations of each assessment technique can be found in [74–77]. It can be seen from the table that, currently, no single technique can effectively identify all potential problems of adhesively bonded FRP-wood structures under service conditions. A combination of several techniques might be useful for a better understanding of the potential problems of such bonded structures throughout their life spans. Thus, there is an urgent need to develop, in future research, more advanced and effective in situ assessment techniques, which will help for easier and better evaluation of the bond performance of adhesively bonded FRP-reinforced wood systems.

Table 4 A summary of NDT and SDT techniques for the assessment of FRP-wood structures

Method	Locate deterioration	Quantify deterioration	Assess strength	Determine stiffness	Identify hidden details
Visual inspection	Limited				
Remote visual inspection	Limited	Limited			Yes
Acoustic emission	Yes				Limited
Digital radioscopy	Yes	Limited			Yes
Infrared thermography	Yes	Limited			Limited
Stress waves	Limited	Limited	Limited	Estimate	
Glueline test	Limited		Limited		
Pin pushing	Yes	Limited	Estimate		
Vibrational analysis	Yes	Limited	Limited	Limited	

6 Outlook

It is clear from the discussions above that using FRP composites in wood reinforcement with appropriate adhesive bonding can effectively improve the stiffness and strength of wood elements. This is beneficial as it increases the application of wood in structures beyond its current limitations and remarkably expands its potential for use in civil infrastructure. The successful implementation of FRP and wood bonding are highly dependent on various factors, from the properties of the initial raw materials to proper surface preparation when implementing the intervention. For each step, the process of application of bonding should follow strictly the specifications to achieve good bonding quality. Currently promising results on adhesively bonded FRP-reinforced wood structures are mainly based on short-term mechanical experiments. The lack of long-term durability data on FRP materials, FRP-wood bond, and even on the structural performance of FRP-reinforced wood systems are essential issues that need to be addressed prior to the wider acceptance of such hybrid wood systems for civil infrastructure application. There is the need for future works on the assessment of bonded FRP-reinforced wood systems under moisture, temperature, fire, and other environmental effects, and under coupled effects and mechanical loading conditions. In addition, simpler, more effective and advanced in situ assessment techniques should be developed to monitor the bond performance of FRP-reinforced wood structures during their service life.

Acknowledgements This work was partially supported by Fachagentur Nachwachsende Rohstoffe e. V. (FNR, Agency for Renewable Resources) funded by Bundesministerium für Ernährung und Landwirtschaft (BMEL, The Federal Ministry of Food and Agriculture of Germany), under the Grant Award No.: 22011617.

References

1. Schober KU, Harte AM, Kliger R, Jockwer R, Xu QF, Chen JF (2015) FRP reinforcement of timber structures. Constr Build Mater 97:106–118
2. Custidio J, Broughton J, Cruz H (2009) A review of factors influencing the durability of structural bonded timber joints. Int J Adhes Adhes 29:173–185
3. Corradi M, Borri A, Castori G, Speranzini E (2016) Fully reversible reinforcement of softwood beams with unbonded composite plates. Compos Struct 149:54–68
4. Lorenzis LVS, La Tegola A (2005) Analytical and experimental study on bonded-in CFRP bars in glulam timber. Compos Part B 36:279–289
5. Fava G, Carvelli V, Poggi C (2013) Pull-out strength of glued-in FRP plates bonded in glulam. Constr Build Mater 43:362–371
6. Yusof A, Saleh AL (2010) Flexural strengthening of timber beams using glass fiber reinforced polymer. Electron J Struct Eng (10). eJSE International:45–56
7. Blass R (2011) Laminated wooden arches reinforced with glass fiber rods. A thesis submitter to the Pennsylvania State University Graduate School in partial fulfilment of the MS degree requirements. Penn State University, State College, PA, 166 p
8. Kasal B, Heiduschke A (2014) Radial reinforcement of glue laminated wood beams with composite materials. Forest Product J 54:74–79

9. Alam P (2004) The reinforcement of timber for structural applications and repair. PhD thesis, University of Bath, 270 p
10. Dias A, Fiorelli J, Molina JC (2015) Numerical analysis of glulam beams without and with GFRP reinforcement. In: 10th international conference on composite science and technology. ICCST/10
11. Fiorelli J, Dias A (2003) Analysis of the strength and stiffness of timber beams reinforced with carbon fiber and glass fiber. Mater Res 6(2003):193–202
12. Ferreira MB, Correa GF, Panzera TH, Fiorelli J, Silva V, Rocco-Lahr R, Christoforo AL (2014) Numerical and experimental evaluation of the use of a glass fiber laminated composite materials as reinforcement in timber beams. Int J Comp Mater 4:73–82
13. Yan LB, Kasal B, Huang L (2016) A review of recent research on the use of cellulosic fibers, their fiber fabric reinforced cementitious, geo-polymer and polymer composites in civil engineering. Compos B Eng 92:94–132
14. Dittenber DB, GangaRao HVS (2012) Critical review of recent publications on use of natural composites in infrastructure. Compos Part A 43:1419–1429
15. Jones MR (1998) Mechanics of composite materials (materials science & engineering series), 2nd edn. CRC Press, 538 p. ISBN-10-156032712X
16. Neitzel M, Mitschang P, Breuer U, Handbuch Verbundwerkstoff. Werkstoffe, Verarbeitung, Anwendung. 2., aktualisierte und erweiterte Auflage. Hanser Verlag, 554 p. ISBN 978-446-43696-1
17. Estrada H, Lee LS (2014) FRP composite constituent mateials. In: Zoghi M (ed) The international handbook of FRP composites in civil engineering. CRC Press, New York, USA, p 32
18. Jawaid HPS, Khahil A (2011) Cellulosic/synthetic fiber reinforced polymer hybrid composites: A review. Carbohyd Polym 86:1–18
19. Yan LB, Chouw N, Jayaraman K (2014) Flax fiber and its composites—a review. Compos B Eng 56:296–317
20. Yan LB, Chouw N, Yuan XW (2012) Improving the mechanical properties of natural fiber fabric reinforced epoxy composites by alkali treatment. J Reinf Plast Compos 31:425–437
21. Prachasaree W, Limkatanyu S, Hat Y (2013) Performance evaluation of FRP reinforced para wood glued laminated beams. Wood Res 58:251–264
22. Gugutsitze G, Draskovic F (2010) Reinforcement of timber beams with carbon fibers reinforced plastics. Slovak J Civil Eng 2:1–6
23. Parvez A, Ansell MP, Smedley D (2009) Mechanical repair of timber beams fractured in flexure using bonded-in reinforcements. Comp Part B 40:95–106
24. Bori A, Corradi M, Grazini A (2005) A method for flexural reinforcement of old wood beams with CFRP materials. Comp Part B 36:143–153
25. Campilho R, de Moura M, Barreto A, Morais J, Domingues J (2010) Experimental and numerical evaluation of composite repairs on wood beams damaged by cross-graining. Constr Build Mater 24:531–537
26. Dourado N, Pereira F, de Moura M, Morais J (2012) Repairing wood beams under bending using carbon–epoxy composites. Eng Struct 34:342–350
27. Franke B, Harte AM (2015) Failure modes and reinforcement techniques for timber beams—state of the art. Constr Build Mater 97:2–13
28. Jia J, Davalos FF (2006) An artificial neural network for the fatigue study of bonded FRP–wood interfaces. Compos Struct 74:106–114
29. Corradi M, Bori A (2007) Fir and chestnut timber beams reinforced with GFRP pultruded elements. Compos Part B 38:172–181
30. Raftery GM, Kelly F (2015) Basalt FRP rods for reinforcement and repair of timber. Compos Part B 70:9–19
31. de la Rosa PG, Escamilla AC, Gonzalez-García MN (2016) Analysis of the flexural stiffness of timber beams reinforced with carbon and basalt composite materials. Compos B 86:152–159
32. Thorhallsson ER, Hinriksson IG, Snæbjornsson JG (2016) Strength and stiffness of glulam beams reinforced with glass and basalt fibers. Compos Part B (in press)

33. Fiore V, Scalici T, Di Bella G, Valenza A (2015) A review on basalt fiber and its composites. Compos B 74:74–94
34. Svecova D, Eden RJ (2004) Flexural and shear strengthening of timber beams using glass fiber reinforced polymer bars-an experimental investigation. Can J Civ Eng 31:45–55
35. Amy K, Svecova D (2004) Strengthening of dapped timber beams using glass fiber reinforced polymer bars. Can J Civ Eng 31:943–955
36. Abejide OS, Ihiabe AV (2016) Probabilistic design of glulam timber beams with bottom CFRP plate. In: Proceedings of the academic conference on agenda for sub-Sahara Africa. Gwagwalada, Abuja FCT-Nigeria
37. Corradi M, Borri A, Righetti L, Speranzini E (2017) Uncertainty analysis of FRP reinforced timber beams. Compos B Eng. https://doi.org/10.1016/j.compositesb.2017.01.030
38. Raftery GM, Harte AM (2013) Nonlinear numerical modelling of FRP reinforced glued laminated timber. Compos Part B 52:40–50
39. Kasal B, Heiduschke A, Kadla J, Haller P (2004) Laminated timber frames with composite fiber reinforced connections. Prog Struct Mat Eng 6:84–93
40. Kasal B, Blass R (2012) Hybrid materials in wood structures—advantages and challenges. An example of reinforcement of a laminated arch. Holztechnologie 53:26–31
41. Kasal B, Pospisil I, Jirovsky I, Drdacky D, Heiduschke A, Haller P (2004) Seismic performance of laminated timber frames with fiber reinforced joints. Earthq Eng Dyn 33:633–646
42. Widmann R, FRP-strengthening of timber. Lecture at ETHZ—HS2015
43. Wood handbook: wood as an engineering material (1999) General technical report FPL-GTR-113. United States Department of Agriculture, Forest Service, Forest Products Laboratory, Madison, WI
44. Azwa NZ, Yousif BF, Manalo AC, Karunasena W (2013) A review on the degradability of polymeric composites based on natural fibers. Mater Des 47:424–442
45. Pizzi A, Mittal KL (2003) Handbook of adhesive technology, 2nd edn. CRC Press, Boca Raton, USA
46. Davalos JF, Qiao P, Trimble BS (2000) Fiber-reinforced composites and wood bonded interfaces: Part I. Durability and shear strength. J Compos Technol Res 22:224–231
47. Lyons JS, Ahmed MR (2005) Factors affecting the bond between polymer composites and wood. J Reinf Plast Compos 24:405–412
48. Valluzzi MR, Nardon F, Garbin E, Panizza M (2016) Multi-scale characterizing of moisture and thermal cycle effects on composite-to-timber strengthening. Constr Build Mater 102:1070–1083
49. Zhou A, Tam LH, Yu Z, Lau D (2015) Effect of moisture on the mechanical properties of CFRP-wood composite: an experimental and atomistic investigation. Compos B Eng 71:63–73
50. Richter K, Steiger R (2005) Thermal stability of wood-wood and wood-FRP bonding with polyurethane and epoxy adhesives. Adv Eng Mater 7:419–426
51. Wang YC, Wong PH, Kodur V (2007) An experimental study of the mechanical properties of fiber reinforced polymer (FRP) and steel reinforcing bars at elevated temperatures. Compos Struct 80:131–140
52. Bai Y, Keller T (2007) Modeling of post-fire stiffness of E-glass fiber-reinforced polyester composites. Compos A 38:2142–2153
53. Italian National Research Council. Guidelines for the design and construction of externally bonded FRP systems for strengthening existing structures—timber structures. CNR-DT 201/2005. Rome, Italy
54. Earnshaw S, Nowicki M, Campbell C (2009) Adhesives for load-bearing timber structures—classification and performance. New Zealand Timber J 17:12–17
55. Zeno M, Tingley D (2006) Fire resistance of FRP reinforced glulam beams. In: Proceedings from the WCTE Portland. Oregon
56. Lartigua J, Coureau JL, Morel S, Galimard P, Maurin E (2015) Effect of temperature on the mechanical performance of guled-in rods in timber structures. Int J Adhes Adhes 57:79–84
57. Tam LH, Zhou A, Yu ZC, Qiu QW, Lau D (2017) Understanding the effect of temperature on the interfacial behavior of CFRP-wood composite via molecular dynamics simulations. Compos B Eng 109:227–237

58. Nardon F, Valluzzi MR, Bertoncello R, Garbin E (2011) Influence of moisture and thermal cycles in bonding of FRP laminates on timber elements. In: International conference on structural health assessment of timber structures 2011. Lisbon, Portugal
59. Korta J, Mlyniec A, Uhl T (2015) Experimental and numerical study on the effect of humidity-temperature cycling on structural multi-material adhesive joints. Compos B Eng 79:621–630
60. Stoeckel F, Konnerth J, Gindl-Altmutter W (2013) Mechanical properties of adhesives of bonding wood—a review. Int J Adhes Adhes 45:32–41
61. Karbhari V, Chin J, Hunston D, Benmokrane B, Juska T, Morgan J et al (2003) Durability gap analysis for fiber-reinforced polymer composites in civil infrastructure. J Compos Const 7:238–247
62. Frangi A, Fontana M, Hugi E, Jübstl R (2009) Experimental analysis of cross-laminated timber panels in fire. Fire Saf J 44:1078–1087
63. Klippel M, Frangi A, Hugi E (2014) Experimental analysis of the fire behavior of finger-jointed timber members. J Struct Eng 140:04013063
64. Zigler R, Pokorny M (2015) Fire protection of timber structures strengthened with FRP materials. Civ Eng J 4:1–8. Article no. 22
65. Ji G, Li G, Li X, Pang SS, Jones R (2008) Experimental study of FRP tube encased concrete cylinders exposed to fire. Compos Struct 85:149–154
66. Williamson TG, Yeh B (2006) Fire performance of FRP reinforced glulam. APA—The Engineered Wood Association. CIB-W18/39-16-1
67. Anon., Low intrusion conservation systems for timber structures (LICONS), http://www.lic ons.org/, 2006
68. CEN TC 193/SC1/WG11 (2007) Adhesives for on site assembling or restoration of timber structures. On site acceptance testing: Part 1: sampling and measurement of the adhesives cure schedule, Doc N20, CEN TC 193
69. CEN TC 193/SC1/WG11 (2007) Adhesives for on site assembling or restoration of timber structures. On site acceptance testing: Part 2: verification of the shear strength of an adhesive joint, Doc N21, CEN TC 193
70. Adams RD, Wake WC (1984) Structural adhesive joints in engineering. Elsevier Applied Science, London and New York
71. Broughton JG, Hutchinson AR (2001) Adhesive systems for structural connections in timber. Int J Adhes Adhes 21:177–186
72. Sernek M, Kamke FE, Glasser WG (2004) Comparative analysis of inactivated wood surface. Holzforschung 58:22–31
73. Nussbaum RM (1999) Natural surface inactivation of Scots pine and Norway spruce evaluated by contact angle measurements. Holz als Roh- und Werkstoff 57:419–424
74. Kasal B, Anthony R (2004) Advances in in situ evaluation of timber structures. Prog Struct Eng Mater 6:94–103
75. Riggio M, Anthony RW, Augelli F, Kasal B, Lechner T, Muller W, Tannert T (2004) In situ assessment of structural timber using non-destructive techniques. Mater Struct 47:749–766
76. Tannert T, Anthony RW, Kasal B, Kloiber M, Piazza M et al (2014) In situ assessment of structural timber using semi-destructive techniques. Mater Struct 47:767–785
77. Kasal B, Tannert T (2011) In situ assessment of structural timber. Springer Verlag, Heidelberg, Germany, p 129
78. DUPONT. Light in weight, high in performance—Kevlar® fiber. http://www.dupont.com/pro ducts-and-services/fabrics-fibers-nonwovens.html

Nanocomposites as Reinforcement for Timber Structural Elements

Clara Bertolini-Cestari and Tanja Marzi

Abstract This chapter presents an overview of recent technological innovations in nanocomposite materials for protection and reinforcement of timber in construction. Starting from the definition of nanotechnologies applied in the field of construction and architectural heritage, the chapter briefly describes nano-materials available in the market. The role of different nano-coatings, their wood surface protection functions, and their compatibility with different wood species are reviewed. On-going experimental research projects for next-generation application fields are presented with a special focus on reinforcement of historic timber joints.

1 Introduction

There have been revolutionary developments in materials science since 1959 when nanotechnologies were first introduced by Richard P. Feynman in his famous lecture, "There's Plenty of Room at the Bottom" [1]. Nowadays, nanotechnology involves multidisciplinary areas of investigation, which in recent years have affected all sectors of the industry with significant economic implications. Nanosciences and nanotechnologies represent a new scientific and technological approach for manipulating material structure and behaviour at the atomic and molecular scales, where properties differ significantly from those observed at a larger scale.

Descending from the normal scale to the infinitely small (the prefix "nano" means 10^{-9}), we enter the domain of quantum physics: the use of nano-particles makes it possible to obtain materials with new chemical, physical and mechanical properties and to increase the original performance of conventional materials (e.g. carbon atoms connected to form nanotubes can produce materials that are stronger than steel) [2].

For the European Union (EU), nanotechnologies represent one of the major fields of scientific development for the near future and have been identified as priority areas for European economic and industrial development. The growing interest in

C. Bertolini-Cestari · T. Marzi (✉)
Department of Architecture and Design, Politecnico di Torino, Turin, Italy
e-mail: tanja.marzi@polito.it

© RILEM 2021
J. Branco et al. (eds.), *Reinforcement of Timber Elements in Existing Structures*,
RILEM State-of-the-Art Reports 33,
https://doi.org/10.1007/978-3-030-67794-7_5

the potential applications of nanotechnology is confirmed at the international level by the strategic documents for the planning of scientific research and technological innovation funding. In 2008, the European Commission adopted the "Code of Conduct for responsible research in the field of nanotechnology." In the absence of comprehensive structured legislation, this voluntary code aims at promoting integrated, safe, ethical and responsible research, recalling some principles (i.e. "sustainability" and "precaution") on which EU states are invited to take concrete action. Being an emerging research field, there is great debate about the extent nanotechnology will benefit or pose risks for human health and environment, and many research projects at the international level are specifically focusing on these issues.

Hundreds of products containing nano-materials are already in use in many sectors (including health, information society, industry, energy, transport and space). In the field of construction and cultural heritage, nanotechnologies are providing a significant boost to innovation in traditional processes and products. Some applications in relevant international architecture contribute to their diffusion. At present, the most promising applications related to nano-structures are coatings. Together with the common performance requirements (long-term stability, durability and weather resistance, good adhesion to the substrate, transparency, sustainability of the production process, etc.), they introduce additional functionality such as self-cleaning, photocatalysis, water resistance, fire resistance, scratch resistance, graffiti resistance and antibacterial properties [3–5].

Currently, the nanotechnology industry is rapidly developing in the field of construction and cultural heritage [6–9]. To date, several ready-to-use nano-products for wood protection are available in the market, but very little research has focused on the reinforcement of timber structures, as most research has been directed at materials such as concrete and metal. The last part of this chapter focuses on this specific topic, presenting also results of experimental research on the application of a polymeric resin reinforced with carbon nanotubes on historic timber structures.

2 Nanotechnology in Timber Reinforcement and Protection

2.1 Overview

Due to its anatomical features, wood is considered a natural nano-structured composite material that is anatomically similar to strong piping and whose cellulose is bonded with a thermoplastic matrix, known as the lignin, and has strong dissipative capacity with regard to fracture energy. It can be seen as a polymeric composite of cellulose, hemicellulose, protein and lignin, as well as, at the nanoscale level, a cellulosic fibrillar composite [10] (Fig. 1). This natural nanocomposite can potentially offer important applications in the field of nanotechnologies, mainly as

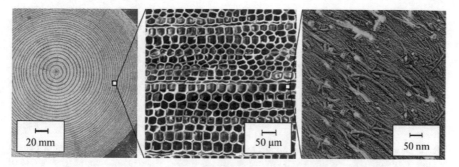

Fig. 1 Variability of wood and its hierarchical structure at different scales of observation [13]

nano-materials derived from forest products (i.e. nano-cellulose, cellulose nanocomposites) or as nanotechnology incorporated into traditional forest-based products (i.e. nano-coating to enhance wood durability) [11, 12].

The use of wood as construction and building material is widespread, with different formal, colour and structural features that make it unique. Architectural heritage built in wood is an important sector of cultural heritage, including its different typologies (such as floors, roofs, bridges, etc.). On this topic, it has become widely accepted that such structures should be maintained and preserved with interventions related to their original conception and material [14, 15]. In the field of conservation of cultural heritage, nanomaterials represent a significant improvement on traditional conservation methodologies, exhibiting the following enhanced properties: (1) they are either non-toxic or they exhibit reduced toxicity compared to traditional restoration materials such as pure solvents; and (2) they allow greater control of restoration intervention (for instance, highly controlled cleaning can be achieved with micro-emulsions and chemical hydrogels, unlike in traditional cleaning methods) [16].

Wood discolours as a result of exposure to ultraviolet (UV) light, moisture and bio-organisms. Treatments introduced by nanotechnologies can improve its durability and stability, which are factors associated with timber exposure to the environment. Despite the innovative (an partially still experimental) character of these treatments, applications of nano-structured materials on relevant new architectures (Fig. 2) [17–20] as well as on historic timber structures belonging to a certain cultural heritage [21–23] was found in recent international case studies.

Nanocomposite coatings can improve the performance and functionality of wood, improving its stability, which is often a factor that limits its use. It should be noted that traditional products and treatments available in the market, aimed at maintaining wood durability, are often highly toxic for humans and for the environment. There is, therefore, the need for new non-toxic products. Nano-materials are usually incorporated into coatings, in an aqueous, organic or polymeric medium, and different nano-impregnation of timber is possible. An essential aspect is that nano-particles must be well dispersed in a suitable medium to avoid aggregation. Nano-particles are less than 100 nm in size and they protect or enhance the properties of the substrate

Fig. 2 Examples of nano-coating application on relevant new architectures: **a** Kamppi Chapel of silence in Helsinki, Finland (K2S Architects, 2012) (photo Marko Huttunen, [17]), **b** phase of application by brush of a pigmented transparent nanotech wax on the façades of spruce wood planks (from [18]), and **c** detail of the façades after two years of exposure showing no change in appearance (courtesy of KS2 Architects, from [19])

(Fig. 3). Their high ratio of surface area to mass ensures that a loading of only a few percent by weight in coatings can significantly enhance chemical, thermal and physical properties [24].

Nanotechnology applications in coatings have shown remarkable growth in recent years. This is as a result of an increased availability of nano-scale materials, such as various types of nanoparticles, and advancements in processes that can control coating structure at the nanoscale [26, 27]. A survey was carried out of nano-based coatings, specifically meant for wood, that are already present in the international market [28, 29]. Among the main fields of application of these products are water repellence and UV ray protection (Fig. 4). It should be highlighted that there is a wide choice of nano-particles which offer different functionality or multi-functionality. Some nano-materials, such as silica (SiO_2) and titania (TiO_2), are multi-functional and there is frequently a choice of nano-materials for a given application.

The increased application of nanotechnology in wood preservation was not preceded by much interest from the scientific community, despite the widespread recognition that nanotechnology has great potential to improve the performance of wood and other cellulosic materials [30, 31].

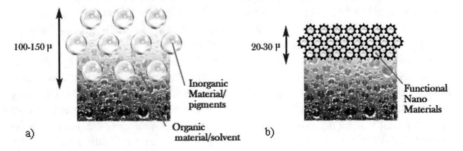

Fig. 3 **a** Traditional paints and coatings are made up of large molecules where water and other organisms can leach into the gaps and erode the surface; **b** Nano-engineered paints are densely packed with robust molecules that act as a penetrative and functional barrier [25]

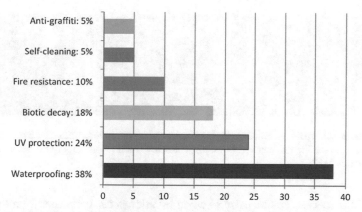

Fig. 4 Main fields of application of nano-based coatings for wood available in the market: waterproofing (Clay, SiO_2, CeO_2, TiO_2); UV protection (TiO_2, ZnO, SiO_2, CeO_2, Fe_2O_3, clays); biotic decay protection (Ag, ZnO, Cu); fire resistance (SiO_2, Clay, TiO_2); self-cleaning/photocatalitic (TiO_2, ZnO); anti-scratching, hardness (TiO_2, Al_2O_3, SiO_2, clays, lime) [28]

Some of the major challenges with respect to nanotechnology-based coatings relate to the dispersion incorporating nanoparticles and the surface characterization that is required in special techniques such as Scanning Probe Microscopy (SPM) (e.g. Atomic Force Microscopy), Scanning Electron Microscopy (SEM), Transmission Electron Microscopy (TEM), X-ray and neutron scattering [26]. Another issue concerns material cost, even if it has come down in recent years mainly due to the increasing number of nanoparticle suppliers, improved manufacturing methods, and increased sales volumes [26].

There are also health and safety issues associated with the abrasion and wear of nano-materials, which leads to the release of potentially hazardous particles into the atmosphere [8, 32]. To understand short- and long-term effects of nanomaterials, in-life and end-of-life health and safety factors should be considered while developing nano-materials for timber [24]. There is the need for further research that would assess and model environmental impacts.

2.2 Fire Resistance

Fire safety is an important consideration in all types of construction, especially when dealing with combustible materials like wood. Nanostructured coatings can be effective as fire retardants. When heated, wood undergoes thermal degradation and combustion to produce gases, vapour, tar and char. In order to improve its resistance to fire, timber can be treated with fire retardants that are typically coated onto the surface (painted, spayed or dipped) or impregnated into the wood structure using vacuum-pressure or other techniques such as plasma treatment [33]. Fire retardants usually provide thermal insulation, absorb the surrounding heat by endothermal reactions, or

Fig. 5 Intumescent paint for wood (right) [37]; tests carried out on wood treated with nanostructured coating to increase fire resistance (top wood); untreated wood element (bottom) before and after fire action (left) [38]

increase the thermal conductivity of wood in order to dissipate heat from the wood surface [34].

The growing awareness of environmental issues and consumer safety of fire retardant products means that there may be a decline in the use of some traditional products (e.g. boron and formaldehyde). The toxicity of fire retardants plays an important role in health and safety legislation [35].

A wide range of fire retardant treatment systems for wood, including nanocomposites and layer by layer applications, have been studied in recent years, and many others are under development (Fig. 5). Fire resistance can be increased by the use of high-performance, thin-film fire-retardant coatings based on the nanoparticles of titanium dioxide and silicon dioxide. These products, which may be sprayed using standard airless equipment, have been formulated to retard the spread of flame across a wide variety of materials and also to suppress the generation of smoke. In the event of fire, they produce water and gases, which snuff out oxygen and produce a cooling effect at the flame front. Dense char is formed, which further protects the surface from combustion [36].

Other studies have addressed wood-based materials, such as research on the treatment of natural wood veneers with nano-oxides (SiO_2, TiO_2 and ZrO_2) to improve their fire behaviour [39]. A mixture of siliceous and phosphate salts has also been used to protect wood from the effect of fire. Due to its fire-retardant properties, it can potentially protect any solid organic compound by absorbing oxygen when the fire starts to generate heat. The coating initially retards the combustion of the organic solid, and when the temperature rises above 600 °C, it carbonises without producing any flame [40, 41].

The high thermal conductivity of nano-silver coatings has also been tested to improve heat transfer in wood and enhance fire resistance [35]. Nano-silver treatment has shown potential to improve some fire-retarding properties of solid wood products. Such coatings may delay thermal degradation and carbonisation by reducing the accumulation of heat through the transfer of heat. Wollastonite nano-fibres were also reported to substantially improve the durability of poplar wood and the fire-retarding properties of solid woods and wood composite materials [42, 43]. Other important aspects of fire-retardant nano-coatings should still be studied in detail, such as the toxicity of the emitted gases and smoke as well as hygroscopicity.

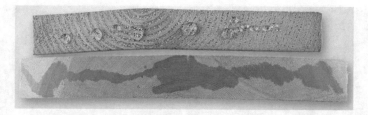

Fig. 6 Waterproofing effect of a timber surface treated with a nano-coating based on silica nanoparticles (top) and untreated wood (bottom) [44]

2.3 Waterproofing

Hydrophobic (water-repellent) nano-coatings are among the promising developments for façade treatments and weather protection of timber. These products mainly consist of silica nanoparticles. By exploiting the so-called "lotus effect", properties of water repellence and, at the same time, breathability of the surface can be obtained (Fig. 6). A hydrophobic nanostructured coating, which is a nearly volatile organic compound (VOC), free and based on a water-borne silane system, has good water repellent functions but it is open-vapour, which is an essential property of wood.

2.4 UV Protection

Ultraviolet rays can change the structure of wood in the sub-surface layer, causing lignin degradation and discoloration. A nanoscale protective barrier can increase resistance to ultraviolet radiation, preserving the appearance and the original colours of the wood, which can be very important when dealing with cultural heritage. The small size of the particles makes it possible to offer high protection without affecting the transparency of the coating [22]. By using this technology, the product reflects the most damaging waves of the sun's spectrum, protecting the wood for a longer time from solar radiation (Fig. 7).

Furthermore, nanoscale solutions are able to concentrate more active substances in a smaller volume of liquid. Therefore, it is possible to use less quantity of the product to achieve durability.

The use of nanoparticles to improve the performance of pigmented and transparent coatings on wood was recently reviewed [46]. The review focused on the effects of nanoparticles on the UV absorption characteristics and physical properties of such coatings. The authors concluded from the literature they reviewed that inorganic nanoparticles, depending on the type, loading, size and dispersion, can be efficient UV absorbers in coatings [46]. Almost all of the recent studies that tested the durability of coatings containing nanoparticles employed artificial weathering [47], but natural weathering trials are needed to fully test the effectiveness of inorganic nanoparticles at improving coating performance [46].

Fig. 7 Durability tests of wooden boards impregnated with UV-protected nano-coatings and unprotected boards [45]

Nano-scale titanium dioxide is a well-known catalyst and photocatalytic material. It has a high refractive index and strong UV light-absorbing capabilities, which help prevent light-induced weathering of the underlying wood [48]. The associated literature is extensive [32, 48–50], analysing its applications as a self-cleaning, UV-absorbing, or sterilising agent. A key factor is that titania is not degraded by oxidative reactions and its photocatalytic action is perpetuated. Furthermore, nano-titania has a very high ratio of surface area to weight, which enhances its photocatalytic action.

Recent research has examined the photochemical stability of water-based acrylic paints containing anatase and rutile nano-titania [51]. Research carried out, using zinc oxide, silica, alumina and titanium dioxide active nano-particles dispersed in acrylic-based, waterborne, solid-colour stains for exterior wood [52–54], to investigate the properties of nano-coatings found significant improvement in UV-shielding of the nanocomposite coatings, demonstrated through accelerated weathering.

The durability of the wood surfaces of different wood species, typical of the Alpine area, exposed to extreme outdoor conditions (Fig. 8) has been investigated [55]. Outdoor wooden furniture was coated with traditional UV-absorbent coatings and nano-coatings, with the objective to assess which of the two required the minimal maintenance. Natural weathering tests were carried out during outdoor exposure in the mountains for a period of two years [56]. Best results were achieved with a polyurethane resin modified with silica nanoparticles from sol-gel [57].

Some studies have focused on the use of tourmaline nanoparticles (a semi-fluid phase of nano- and micro-sized particle elbaite). Their main action consists in the emission of negative ions and far infrared radiation (FIR) which induces self-cleaning, ionising and bactericidal effects. The incorporation of tourmaline nanoparticles in silver nanoparticles was found to increase the conductivity of the solution, resulting in improved tensile strength and high bactericidal activity [40, 41, 58] of the coating.

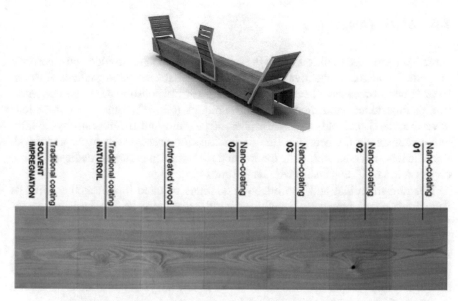

Fig. 8 Durability test of larch wood used for outdoor furniture in the Alpine area impregnated with different UV absorber coatings (2 traditional and 4 nano-coatings). In the image the difference after 2 months of outdoor exposure is visible [55]

2.5 Abrasion Resistance

To protect the environment and ensure the safety of workers, there is increased application of bio-based coatings made of renewable resources, such as oil finishes, in the wood coating market. However, vegetable oil-based finishes do not provide the scratch and wear resistance required, for example, in flooring applications [59]. The addition of nanoparticles, like alumina, silica and zirconia, has been reported to improve the scratch and wear resistance of coatings [26, 60, 61]. A specific study was carried out to improve the performance of a vegetable oil-based coating by the addition of nanoparticles (nanoclay, nanosilica, nanoalumina). Homogeneous dispersion of nanoparticles is a crucial point in the determination of the performance of coatings. Therefore, several mixing techniques have been tested to obtain the best results for a bio-based coating using a high-speed mixer with the addition of glass beads. The addition of 1% nanoparticles significantly increased the abrasion resistance of a modified coating after 2000 abrasion cycles in comparison with a pure unmodified coating (based on natural resins, vegetable wax and refined solvent), as applied by brush on larch wood flooring samples [59]. Recent work at the University of Bath has examined the effect of the application of nano-lime to the end-grain of Welsh woods, which are used in flooring applications to improve indentation hardness [24].

2.6 Biotic Decay

Lime has been used traditionally in the treatment of wood to provide anti-microbial properties. It acts as a reflective coating and enhances the aesthetic qualities of wood. Even if nano-limes have been studied mainly for the consolidation and conservation of limestones and lime-based wall paintings [62, 63], some researchers [64] have examined the role of nanoparticles of calcium and magnesium hydroxides and carbonates in the neutralisation of acidification processes in paper and wood. Specific research concerning the treatment of the wooden parts was dedicated to the conservation of *Vasa* warship with nano-limes [65].

The anti-microbial and antireflective qualities of nano-lime, together with its deacidifying and breathable qualities, could potentially be the basis of multifunctional coatings for wood [24].

Improved protection from biotic decay of wood can be obtained by using nanomaterials in protective treatments and as a barrier to moisture. In particular, products containing silver nanoparticles are effective for their antibacterial, anti-microbial and anti-mould properties, and nano-silver is nowadays a constituent of several commercial wood preservatives (Fig. 9).

Wood can be attacked by a variety of insects, fungi and other organisms. These decay agents can be tackled by using preparations based on nano-silver. Specimens coated with nano-silver leave active nanoparticles on the surface. When bacteria come in contact with the protected surface, the silver interacts with the bacteria and causes their enzymes to break down, stopping their activities and causing them to die. Silver is a well-known biocidal agent and it has been available for over 100 years in a colloidal nano-form [66, 67].

Silver-containing compounds, which release silver ions into moist environment, differ from nano-silver particles, which may act as a catalyst and are often held in a polymeric matrix as a coating. The release of nano-silver by leaching from paints

Fig. 9 Wooden façade protected with a nanocoating (lower part); unprotected (upper part) façade showing the presence of mould (left) [21]; durability test carried out on wood composite panels protected and unprotected (right) with a nanosilver coating [22]

[68] is often associated with conversion into less toxic compounds such as silver sulphide, so health risks are reduced.

Nano-silver has been widely utilised to reduce the biodegradation of wood and also to improve wood resistance to termite attack [36, 69].

The resistance of Scots pine wood, vacuum-treated with nano-particles of some metal oxides such as ZnO, B_2O_3, CuO, TiO_2, CeO_2, and SnO_2, to decay, leachability, mould, fungi and subterranean termites was evaluated in a recent study [70].

A study aimed at enhancing the performance of existing wood coatings by the addition of nanoparticles [41]. Multi-functional nanocomposite materials have been tested on historical timber elements to obtain a coating with reduced environmental impact, leading to decreased costs of maintenance of timber structures. Samples of different wood species (oak, chestnut and beech) belonging to the architectural heritage of the Piedmont Region have been treated with coatings containing different nanomaterials (nano-silver, nano-tourmaline, nano silica phosphoric salts, titanium dioxide). The aim was to obtain a product easy to apply on site, with waterproofing and fire-resistance properties, and that provides protection from biotic decay. Different physico-chemical analyses, as well as radiometric analysis, have been carried out (Fig. 10).

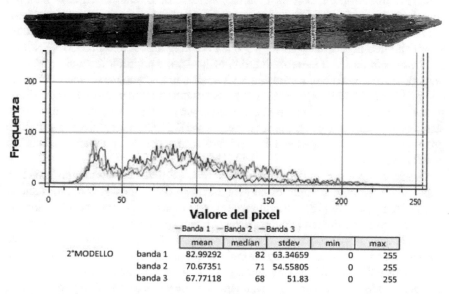

		mean	median	stdev	min	max
2°MODELLO	banda 1	82.99292	82	63.34659	0	255
	banda 2	70.67351	71	54.55805	0	255
	banda 3	67.77118	68	51.83	0	255

Fig. 10 Radiometric analysis of a timber element belonging to a historic timber structure of the 17th century treated with different nano-coatings containing nano-silver and nano-tourmaline. Analysis of the histograms for the entire orthophotos (true colour RGB) for photogrammetric 3D models after treatment, showing the distribution of the radiometric values in the three bands of the image [41]

3 Reinforcement

The application of composite nano-materials as reinforcement for timber structures is not yet widespread. Some research has been conducted in recent years on the use of nanotechnology in new wood constructions [71]. Other research assessed the in-situ bonding of steel or composite pultruded rods and plates and fibre-reinforced polymers (FRPs) for the repair and strengthening of timber structures by epoxy adhesive modified with different nano- and micro-particles [72, 73]. Structural adhesives are designed to support loads, are often subject to severe environmental conditions, and are required to possess excellent mechanical and thermal properties. The addition of nanofillers/nano-reinforcement into thermosets such as epoxy has been recently tested. Nano-silica, nano-rubber and ceramic micro-particles were added to control rheology and improve structural adhesive bonding. The adhesives were back-injected into oversize holes or slots and rods or plates were pushed into the adhesives to form a void-free interface, after which the adhesives were allowed to cure at room temperature. The bond integrity and strength depends on a number of factors, including the type of adhesive and adherence, cure cycle, bond-line thickness, viscosity, and thermal properties [72, 74]. The addition of nano- and micro-fillers, which are thixotropic and shear-thinning, increased the bond strength of the base adhesive by up to 20%. However, the measured contact angle negatively correlated with the measured bond strength. Overall, the less ductile, micro-particle-filled adhesives exhibited higher shear strengths when bonding timber and FRP rods [72]. As for the effects on the long-term duration of loads, the addition of liquid rubber as nano-fillers to a base commercial adhesive was found to induce strain recovery under constant load, leading to an increased glass transition temperature of the modified product, a reduced initial creep rate in the primary creep stage, and a much more stable steady-state creep in the second creep stage [75]. Currently only a few of these products are commercially available and further research and product development is still required.

Other research focused on the use of nanoparticles of aluminum oxide to improve the bonding strength of polyvinyl acetate (PVA) in wet conditions and at elevated temperatures [76]. Aluminum oxide nano-particles are one of the major and frequently used engineered nanoparticles, known for their hardness and excellent scratch and abrasion properties. The shear strengths of wood joints were found to be improved, under all conditions, by the inclusion of nanoparticles to PVA. The thermal stability of PVA was affected by nanoparticles as well. Nanostructure studies have shown that adequate dispersal of nanoparticles in PVA is a crucial step to obtaining nanocomposites with superior properties. Over the past decades, nanoparticles have opened up opportunities for polymer industries to improve the properties of polymers beyond the limits reachable by the addition of micro-particles. Polymer-based nanocomposites exhibit much better mechanical, thermal and barrier properties than composites made from polymers and their corresponding conventional fillers [76].

The addition of nanoparticles to polymers has no negative effects on polymer properties, and often a drastic improvement in the properties can be obtained by the addition of only a small amount, even as low as 1%.

In a recent study [77], a critical analysis of the currently employed wood consolidation products was performed and the opportunities offered by some nano-insertions were investigated. The use of nano-materials and composite materials to strengthen wood has been suggested by several researchers who studied the effects of carbon nanotube-based composites on the mechanical characteristics of reinforced wooden elements [71]. In Italy, given the greater relevance of rehabilitation and conservation, more research has focused on the methodologies for consolidating existing historical structures [78–80]. Experiments carried out on ancient timber structures are very complicated, since the mechanical characteristics of timber vary on account of decay and deterioration. However, experimental results have shown a clear trend in terms of efficiency concerning the mentioned rehabilitation techniques [81].

The development of polymers reinforced with nanoparticles is one of the most promising approaches in the field of future engineering applications. The unique properties of some nanoparticles (carbon black [71] and carbon nanotubes) and the possibility of combining them with traditional reinforcing elements (glass fibre, carbon fibre or Kevlar™) have generated an intense research interest in the field of nanocomposites [71, 82, 83]. Carbon black is made up of particles with a diameter of 30 nm. It is commonly used to make polymers conductive and prevent their accumulation of electric charges. Carbon nanotubes (CNTs) have diameters of several nanometres (1–50 nm) and lengths of several microns (up to 10 μm); they have very good potential to improve the electrical and mechanical properties of polymers, even with an addition of 0.1% weight content, compared to epoxy resin [71, 84]. Transferring their remarkable mechanical, thermal and electrical properties to the polymer matrix is, however, not easy. Consequently, the correct dispersion of CNTs in a polymer and timber interface is a crucial factor [71, 85]. The difficulty in dispersing CNTs arises from their inherently hydrophobic nature and tendency to agglomerate due to their size and shape [85]. Moreover, with epoxy resin matrices, the issue of dispersion is further complicated considering that they are different from common polymers, as most epoxy resins have two components, a resin and a hardener.

Different techniques for dispersing CNTs in solvents (acetone and ethanol), in addition to ultrasound, mechanical shaking and a combination of the two techniques, have already been used [71, 86]. The interface adhesion can be improved by chemically functionalizing the CNT surfaces; this generates strong covalent-type bonding [82, 83, 87–89]. The bonding between carbon nanotubes and polymers allows the strain to be transferred from one phase to the other. Double wall carbon nanotubes (DWCNT) with or without a functional group surface can be purchased in the market. There are some experimental difficulties, as seen in the nature of the results obtained with epoxy resins, where, by adding single wall carbon nanotubes (SWCNT) in the best cases, the bending strength is slightly increased, although in most cases, the bending strength is often reduced [88]. Recently, fabrics containing up to 39% in SWCNT weight impregnated with epoxy resin have been produced; once again, the results obtained fell below expectations [71]. On the other hand, significant progress in mechanical resistance values has been made with polymethyl methacrylate (PMMA), polyvinylic alcohols and polystyrene-based composites [88]. Promising results have also been achieved with polyurethane resin composites, with up to 10% in weight of carbon nanotubes [90].

Referring to the experimentation on wood, CNTs potentially provide a number of advantages:

- They are morphologically and chemically compatible with polymer resins used as bonding materials and with wood. This is due to the fact that CNTs are anatomically similar to strong pipes bonded with a thermoplastic matrix and which have strong dissipative capacity with regard to fracturing energy;
- They allow the polymer bonding matrix to improve its own inbuilt deformation capacity;
- Resin-fibre compounds have excellent mechanical characteristics due to the high specific resistance of the fibres, which have great cohesive strength combined with high ductility. These characteristics ensure significant creep resistance within the composite;
- The tubular structure of nanofibres has great permeability to vapour: this is a crucial characteristic when dealing with large surfaces treated with glue, especially in the case of wood, since any accumulation of moisture must be easy to disperse to avoid biotic degradation.

Polymeric resin reinforced with carbon nanotubes (CNTs) was applied for localised interventions on historic timber structures [28, 78, 79]. The study was concerned with the conservation of a certain wooden joint belonging to traditional constructions and attempted to define and assesse a methodology of preparation and application of the nanocomposite. The aim was to verify whether the mechanical resistance increased in comparison with traditional reinforcement methods. Different wood species were considered and the experimentation was carried out with a view to obtaining possible in-situ applications of the technique. Laboratory tests were carried out first on small wooden samples, and afterwards on full-scale wooden elements (Fig. 11) [80]. The objective of the study was to obtain a nanomaterial, based on polymeric resin reinforced with carbon nanotubes, for the reinforcement of timber structures. The new material, in principle, could be used as an adhesive, a surface consolidant (like in the present case) or as a deep-impregnation consolidant in combination with vacuum techniques [91].

4   Conclusions and Outlook

Wood, as a natural nanocomposite, can potentially offer important applications in the field of nanotechnologies. Besides the commonly required performance (long-term stability, durability, weather resistance, good adhesion to the substrate, transparency, sustainability for the production process, etc.), it is possible, through

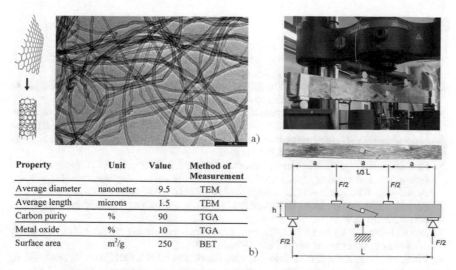

Property	Unit	Value	Method of Measurement
Average diameter	nanometer	9.5	TEM
Average length	microns	1.5	TEM
Carbon purity	%	90	TGA
Metal oxide	%	10	TGA
Surface area	m²/g	250	BET

Fig. 11 TEM image (**a**) and CNT characteristics of Nanocyl© series 7000 (**b**) [92]. Decayed wooden samples; bending tests and scheme for the mechanical resistance of timber "Jupiter" joints (L = 1000 mm, h = 100 mm, B = 35 mm) [28]

nano-structured coatings, to introduce additional surface functionality such as self-cleaning, photocatalysis, water resistance, fire resistance, scratch resistance and antibacterial functionality.

The main nano-materials available in the market, nano-coatings and their wood surface protection functions, their classification, and their compatibility with wood have been reviewed in this chapter. On-going experimental research projects for potential new fields of application in the sectors of architecture, civil engineering and cultural heritage have also been analysed, with a special focus on the reinforcement of historical timber structures. The results of an experimental research activity on the application of polymeric resin reinforced with CNTs on historical timber structures are promising and constitute a base for future developments in the application of CNTs.

Long-term research to assess and model environmental impacts and the economic viability of using nano-scale materials in wood surfaces and timber structures is still needed for most of the applications. There are clear benefits associated with the adoption of many nano-enabled products, but there may also be hazards, such as potential health risks, associated with some nanomaterials. The exact form and dose of the materials involved are important determinants of the significance of the risk. Thus, a good understanding of the materials being used [27] and of their behaviour over time is important. Until the health and environmental hazards of these materials are clearly assessed, many of their applications will remain at an experimental level.

References

1. R. Feynman, there's plenty of room at the bottom, American Physical Society, California Institute of Technology (1959) Eng Sci 23(1960):22–36
2. Ashby MF, Ferreira PJ, Schodek DL (2009) Nanomaterials, nanotechnologies and design. An introduction for engineers and architects. Elsevier, Paris
3. Bartos PJM, De Miguel Y, Porro A (eds) (2005) Proceedings of the 2nd international symposium on nanotechnology in construction, 13–16 November 2005, Bilbao (Spain), RILEM, Bagneux
4. Elvin G (2007) The nano revolution. A science that works on the molecular scale is set to transform the way we build, Architect Magazine
5. Hincapié I, Künniger T, Hischier R, Cervellati D, Nowack B, Som C (2015) Nanoparticles in facade coatings: a survey of industrial experts on functional and environmental benefits and challenger. J Nanopart Res 17:287. https://doi.org/10.1007/s11051-015-3085-3
6. Blee A, Matisons JG (2008) Nanoparticles and the conservation of cultural heritage. Mater Forum 32:121–128
7. Zhu W, Bartos PJM, Porro A (2004) Application of nanotechnology in construction. Summary of a state-of-the-art report. Mater Struct 37:649–658
8. van Broekhuizen P, van Broekhuizen F, Cornelissen R, Reijnders L (2011) Use of nanomaterials in the European construction industry and some occupational health aspects thereof. J Nanopart Res 13:447–462. https://doi.org/10.1007/s11051-010-0195-9
9. Das BB, Mitra A (2014) Nanomaterials for construction engineering-a review. Int J Mater Mech Manuf 2(1):41–46. https://doi.org/10.7763/IJMMM.2014.V2.96
10. AAVV (2004) Nanotechnology for the forest products industry. Vision and Technology Roadmap, Lansdowne, U.S.A., October 17–19. TAPPI Press, Atlanta
11. Cai Z, Rudie A, Stark N, Sabo R, Ralph S (2013) New products and product categories in the global forest sector. In: Hansen E, Panear R, Vlosky R (eds) The global forest sector: changes, practices, and prospects. CRC Press, Boca Raton, pp 129–149
12. Jacoby M (2014) Nano from the forest. Tiny cellulosic particles are poised to make big impact on materials technology. C&EN Chem Eng News 92(26):9–12
13. De Borst K, Jenkel C, Montero C, Colmars J, Gril J, Kaliske M, Eberhardsteiner J (2013) Mechanical characterization of wood: an integrative approach ranging from nanoscale to structure. Comput Struct 127:53–67. https://doi.org/10.1016/j.compstruc.2012.11.019
14. Bertolini Cestari C, Marzi T, Seip E, Touliatos P (eds) (2004) Interaction between science, technology and architecture in timber construction. Elsevier, Paris
15. Kasal B, Tannert T (2010) In Situ assessment of structural timber. Springer. https://doi.org/10.1007/978-94-007-0560-9
16. Baglioni P, Chelazzi D, Giorni R (2015) Nanotechnologies in the conservation of cultural heritage. A compendium of materials and techniques. Springer. https://doi.org/10.1007/978-94/017-9303-2
17. Lintula K, Sirola N, Summanen M (2012) Kamppi chapel of silence. PUU 2:6–13
18. https://learnsee.wordpress.com/2012/07/29/kamppi-chapel-finishing-touch/. Date of access January 2017
19. http://www.architectmagazine.com/technology/detail/a-curved-wooden-chapel-in-a-northern-square_o
20. Servais F (2013) Rénovation et extension de l'Hôtel du Val d'Amblève à Stavelot, Les Cahiers nouveaux N° 87, pp 99–100
21. The NanoPhos Casebook: applications from across the world, Jan 2013. http://www.nanophos.com/en/. Date of access October 2014
22. www.nanobiz.com. Date of access October 2014
23. Vita I, Fernandez F, Scognamiglio M, Bellanca L (2009) Applicazione di prodotti nanostrutturati per la protezione di superfici storiche nella Cappella Palatina (Pa). In: Sposito A (ed) Nanotech for architecture. Luciano Editore, pp 369–380
24. Ansell MP (2013) Multi-functional nano-materials for timber in construction. Proc ICE Const Mater 166(4):248–256. https://doi.org/10.1680/coma.12.00035

25. http://www.icannanopaints.com/nanoTechnology.html. Date of access May 2015
26. Fernando RH (2009) Nanocomposite and nanostructured coatings: recent advancements. In: Fernando R et al (eds) Nanotechnology applications in coatings; ACS symposium series. American Chemical Society, Washington, DC, pp 2–21. http://dx.doi.org/10.1021/bk-2009-1008.ch001
27. Jones W, Gibb A, Goodier C, Burst P, Song M, Jin J (2016) Nanomaterials in construction—what is being used, and where? In: Proceedings of the ICE, construction materials, pp 1–14. http://dx.doi.org/10.1680/jcoma.16.00011
28. Marzi T (2010) Impiego di nanotecnologie nei beni culturali per l'efficienza di sistemi manutentivi del costruito in legno: tecnologie innovative di recupero (Nanotechnologies/nanosciences: from wood improvement to the reinforcement of timber and monitoring of interventions). PhD Thesis in Innovation Technology for Built Environment, tutor Prof. Bertolini Cestari C., Politecnico di Torino
29. Marzi T (2015) Nanotechnologies for reinforcement and protection of timber structures: a review and future challenges. Constr Build Mater 97:119–130. https://doi.org/10.1016/j.conbuildmat.2015.07.016
30. Vlosky RP (2009) Statistical overview of the U.S. Wood Preserving Industry: 2007. Louisiana Forest Products Development Center
31. Evans PD, Matsunaga H, Kiguchi M (2008) Large-scale application of nanotechnology for wood protection. Nat Nanotechnol 3(10):577. https://doi.org/10.1038/nnano.2008.286
32. Nowack B, Bucheli TD (2007) Occurrence, behavior and effects of nanoparticles in the environment. Environ Pollut 150(1):5–22. https://doi.org/10.1016/j.envpol.2007.06.006
33. Blanchet P, Landry V (2015) Nanocomposite coatings and plasma treatments for wood-based products. In: Ansell M (ed) Wood composites. Elsevier. ISBN: 978-1-78242-454-3. http://dx.doi.org/10.1016/B978-1-78242-454-3.00013-5
34. Lowden LA, Hull TR (2013) Flammability behaviour of wood and a review of the methods for its reduction. Fire Sci Rev 2:4. https://doi.org/10.1186/2193-0414-2-4
35. Taghiyari HR (2012) Fire-retarding properties of nano-silver in solid wood. Wood Sci Technol 46:939–952. https://doi.org/10.1007/s00226-011-0455-6
36. Bertolini C, Crivellaro A, Marciniak M, Marzi T, Socha M (2010) Nanostructured materials for durability and restoration of wooden surfaces in architecture and civil engineering. In: Ceccotti A, Wan De Kuilen J-W (eds) Proceedings of 11th world conference on timber engineering, vol III, pp 705–706
37. http://www.coating.co.uk/intumescent-paint-for-wood/. Date of access January 2017
38. www.percenta.com. Date of access October 2014
39. Francés Bueno AB, Navarro Bañón MV, Martínez de Morentín L, Moratalla García J (2014) Treatment of natural wood veneers with nano-oxides to improve their fire behaviour. In: 2nd international conference on structural nano composites (NANOSTRUC 2014). IOP conf. series: materials science and engineering, vol 64, p 012021. IOP Publishing. https://doi.org/10.1088/1757-899x/64/1/012021
40. www.nctchemical.com/en. Date of access May 2016
41. Spanò A, Bertolini Cestari C, Marzi T, Maspoli R, Torretta A, Percivalle E, Nicola M (2016) Final report of the regional feasibility study "WOOD_defender", Politecnico di Torino
42. Haghighi Poshtiri A, Taghiyari HR, Karimi AN (2014) Fire-retarding properties of nano-wollastonite in solid wood. Philip Agric Sci 97(1):52–59
43. Taghiyari HR (2014) Nanotechnology in wood and wood-composite materials. J Nanomater Mol Nanotechnol 3:1. https://doi.org/10.4172/2324-8777.1000e106
44. https://grasi.en.alibaba.com/. Date of access January 2017
45. http://www.vistapaint.fr. Date of access January 2017
46. Evans PD, Haase JG, Shakri A, Semanand BM, Kiguchi M (2015) The search for durable exterior clear coatings for wood, coatings. vol 5, pp 830–864. https://doi.org/10.3390/coatings5040830
47. Nikolic M, Lawther JM, Sanadi AR (2015) Use of nanofillers in wood coatings: a scientific review. J Coat Technol Res 12:445–461

48. Srinivas K, Pandey KK (2017) Enhancing photostability of wood coatings using titanium dioxide nanoparticles. In: Pandey K, Ramakantha V, Chauhan S, Arun Kumar AN (eds) Wood is good. Current trends and future prospects in wood utilization. Springer, Singapore, pp 251–250. https://doi.org/10.1007/978-981-10-3115-1
49. Auclair N, Riedl B, Blanchard V, Blanchet P (2011) Improvement of photoprotection of wood coatings by using inorganic nanoparticles as ultraviolet absorbers. Forest Prod J 61(1):20–27
50. Blanchard V, Blanchet P (2011) Color stability for wood products during use: effects of inorganic nanoparticles. BioResources 6(2):1219–1229
51. Allen NS, Edge M, Ortega A et al (2002) Behaviour of nanoparticle (ultrafine) titanium dioxide pigments and stabilisers on the photooxidative stability of water based acrylic and isocyanate based acrylic coatings. Polym Degrad Stab 78(3):467–478. https://doi.org/10.1016/S0141-391 0(02)00189-1
52. Vlad-Cristea M, Riedl B, Blanchet P (2010) Enhancing the performance of exterior waterborne coatings for wood by inorganic nanosized UV absorbers. Prog Org Coat 69:432–441. https://doi.org/10.1016/j.porgcoat.2010.08.006
53. Vlad-Cristea M, Riedl B, Blanchet P (2011) Effect of addition of nanosized UV absorbers on the physicomechanical and thermal properties of an exterior waterborne stain for wood. Prog Org Coat 72:755–762. https://doi.org/10.1016/j.porgcoat.2011.08.007
54. Vlad-Cristea M, Riedl B, Blanchet P, Jimenez-Pique E (2012) Nanocharacterization techniques for investigating the durability of wood coatings. Eur Polym J 48:441–453. https://doi.org/10.1016/j.eurpolymj.2011.12.002
55. Germak C, Bozzola M (2013) Technical report of the EU project Alcotra 2007 "Savoir Bois". Politecnico di Torino
56. Germak C, Bozzola M (2013) Savoir bois: culture and furniture for the mountain. Paesaggio Urbano 4:42–47
57. Raghavan S, Gopagani R, Kalidindi R (2013) Multifunctional sol–gel coatings for protection of wood. Wood Mat Sci Eng 8(4):226–233. https://doi.org/10.1080/17480272.2013.834967
58. Tijing IP, Amarjargal A, Jiang Z, Ruelo M, Park C-H, Pant HR, Kim D-W, Lee DH, Kim CS (2013) Antibacterial tourmaline nanoparticles/polyurethane hybrid mat decorated with silver nanoparticles prepared by elecrospinning and UV photoreduction. Curr Appl Phys 13(1):205–210. https://doi.org/10.1016/j.cop.2012.07.011
59. Nejad M, Cooper P, Landry V, Blanchet P, Koubaa A (2015) Studying dispersion quality of nanoparticles into a bio-based coating. Org Coat 89:246–251. https://doi.org/10.1016/j.porgcoat.2015.09.006
60. Landry V, Blanchet P, Riedl B (2010) Mechanical and optical properties of clay-based nanocomposites coatings for wood flooring. Prog Org Coat 67(4):381–388
61. Amerio E et al (2008) Scratch resistance of nano-silica reinforced acrylic coatings. Prog Org Coat 62(2):129–133
62. Dei L, Salvadori B (2006) Nanotechnology in cultural heritage conservation: nanometric slaked lime saves architectonic and artistic surfaces from decay. J Cult Herit 7(2):110–115. https://doi.org/10.1016/j.culher.2006.02.001
63. Daniele V, Taglieri G, Quaresima R (2008) The nanolimes in cultural heritage conservation: characterisation and analysis of the carbonatation process. J Cult Herit 9(3):294–301. https://doi.org/10.1016/j.culher.2007.10.007
64. Baglioni P, Giorgi R (2006) Soft and hard nanomaterials for restoration and conservation of cultural heritage. Soft Matter 2(4):293–303. https://doi.org/10.1039/B516442G
65. Baglioni P, Chelazzi D, Giorgi R (2005) Nanoparticles of calcium hydroxide for wood conservation, The deacidification of the Vasa warship. Langmuir 21:10743–10748
66. Nowack B, Krug HF, Height M (2011) 120 years of nanosilver history: implications for policy makers. Environ Sci Technol 45(4):1177–1183
67. Rai M, Yadav A, Gade A (2009) Silver nanoparticles as a new generation of antimicrobials. Biotechnol Adv 27(1):76–83. https://doi.org/10.1016/j.biotechadv.2008.09.002
68. Kaegi R, Sinnet B, Zuleeg S et al (2010) Release of silver nanoparticles from outdoor facades. Environ Pollut 158(9):2900–2905. https://doi.org/10.1016/j.envpol.2010.06.009

69. Kartal SN, Green F, Clausen CA (2009) Do the unique properties of nano-metals affect leachability or efficacy against fungi and termites? Int Biodeterior Biodegrad 63(6):490–495
70. Terzi E, Nami Kartal S, Nural Yılgör N, Rautkari L, Yoshimura T (2016) Role of various nano-particles in prevention of fungal decay, mold growth and termite attack in wood, and their effect on weathering properties and water repellency. Int Biodeter Biodegradation 107:77e87. http://dx.doi.org/10.1016/j.ibiod.2015.11.010
71. Breuer O, Sundararaj U (2004) Big return from small fibers: a review of polymer/carbon nanotube composites. Polym Compos 25(6):630–645. https://doi.org/10.1002/pc.20058
72. Ahmad Z, Ansell MP, Smedley D (2010) Epoxy adhesives modified with nano- and micro-particles for in-situ timber bonding: effect of microstructure on bond integrity. Int J Mech Mater Eng 5:59–67
73. Ahmad Z, Ansell MP, Smedley D (2011) Epoxy adhesives modified with nano- and micropar-ticles for in-situ timber bonding: fracture toughness characteristics. J Eng Mater Tech 133:031006-1–031006-9
74. Ahmad Z, Ansell MP, Smedley D, Thermal properties of epoxy-based adhesive reinforced with nano and micro-particles for in-situ timber bonding. Int J Eng Technol IJET-IJENS 10(02):21–27, 104202-6969
75. Pizzo B, Smedley D (2015) Adhesive for on-site bonding: characteristics, testing and prospects. In: Harte AM, Dietsch P (eds) Reinforcement of timber structures. A state-of-the art report. Shaker Verlag, Aachen, pp 113–131
76. Kaboorani A, Riedl B (2012) Nano-aluminum oxide as a reinforcing material for thermoplastic adhesives. J Ind Eng Chem 18:1076–1081. https://doi.org/10.1016/j.jiec.2011.12.001
77. Tuduce-Traˇistaru AA, Câmpean M, Timar MC (2010) Compatibility indicators in developing consolidation materials with nanoparticle insertions for old wooden objects. Int J Conserv Sci 1(4):219–226
78. Bertolini Cestari C, Invernizzi S, Marzi T, Tulliani JM (2008) Nanotechnologies applied to the restoration and maintenance of wooden built heritage. In: D'Ayala D, Forde E (eds) Struc-tural analysis of historical construction. Preserving safety and significance. In: Proceedings of SAHC2008. VI international conference, bath. Taylor & Francis, London, pp 941–947
79. Bertolini Cestari C, Invernizzi S, Marzi T, Tulliani JM (2009) Use of nanotechnologies and nanosciences in cultural heritage for the efficiency of maintenance systems in wooden built heritage: restoration, conservation, maintenance, monitoring of interventions. In: Structures en bois dans le patrimoine bâti, Actes des journées techniques intern. Bois, Les cahiers d'ICOMOS France, Icomos, Paris, pp 87–91
80. Bertolini C, Invernizzi S, Marzi T, Tulliani JM (2013) The reinforcement of ancient timber-joints with carbon nano-composites. Meccanica 48:1925–1935. https://doi.org/10.1007/s11 012-013-9735-6
81. Marzi T (2015) Nanostructured materials for protection and reinforcement of timber structures: innovative nano-coatings. In: Harte AM, Dietsch P (eds) Reinforcement of timber structures. A state-of-the art report. Shaker Verlag, Aachen, pp 209–230
82. Gojny FH, Wichmann MHG, Köpke U, Fiedler B, Schulte K (2004) Carbon nanotube-reinforced epoxy-composites: enhanced stiffness and fracture toughness at low nanotube content. Compos Sci Technol 64:2363–2371. https://doi.org/10.1016/j.compscitech.2004.04.002
83. Lau KT, Lui D (2002) Effectiveness of using carbon nanotubes as nanoreinforcements for advanced composite structures. Carbon 2002:1597–1617. https://doi.org/10.1016/S0008-622 3(02)00157-4
84. Yeh M-K, Hsieh T-H, Tai N-H (2008) Fabrication and mechanical properties of multi-walled carbon nanotubes/epoxy nanocomposites. Mater Sci Eng 483–484:289–292. https://doi.org/10.1016/S0008-6223(02)00133-1
85. Chakraborty AK, Plyhm T, Barbezat M, Necola A, Terrasi GP (2011) Carbon nanotube (CNT)-epoxy nanocomposites: a systematic investigation of CNT dispersion. J Nanopart Res 13:6493–6506. https://doi.org/10.1007/s11051-011-0552-3

86. Chen H, Jacobs O, Wu W, Rüdiger G, Schädel B (2007) Effect of dispersion method on tribological properties of carbon nanotube reinforced epoxy resin composites. Polym Test 26:351–360. https://doi.org/10.1016/j.polymertesting.2006.11.004
87. Miyagawa H, Drzal LT (2004) Thermo-physical impact properties of epoxy nanocomposites reinforced by singlewall carbon nanotubes. Polymer 45:5163–5170. https://doi.org/10.1016/j.polymer.2004.05.036
88. Lau K, Lu M, Lam C, Cheung H, Sheng F, Li H (2005) Thermal and mechanical properties of single-walled carbon nanotube bundle-reinforced epoxy composites: the role of solvent for nanotube dispersion. Compos Sci Technol 65:719–725. https://doi.org/10.1016/j.compscitech.2004.10.005
89. Miyagawa H, Rich MJ, Drzal LT (2006) Thermophysical properties of epoxy nanocomposites reinforced by carbon nanotubes and vapour grown carbon fibers. Thermochim Acta 442:67–73. https://doi.org/10.1016/j.tca.2006.01.016
90. Webster TJ, Waid MC, McKenzie JL, Price RL, Ejiofor JU (2004) Nanobiotechnology: carbon nanofibres as improved neural and orthopaedic implants. Nanotechnology 15:48–54
91. Bertolini Cestari C, Invernizzi S, Marzi T, Tulliani JM (2010) Nano-technologies/smart-materials in timber constructions belonging to cultural heritage. In: Proceedings of 11th world conference on timber engineering, vol IV, pp 761–762
92. http://www.nanocyl.com. Date of access October 2014

Reinforcement of Timber Structures: Standardization Towards a New Section for EC 5

Philipp Dietsch

Abstract The reinforcement of timber structures has seen considerable research and development in recent years. New materials and methods for reinforcement have been developed and are now used in practice. Current design standards, however, lack specific guidance to design reinforcements for timber members and joints. To close this gap in the new generation of Eurocode 5, CEN/TC 250/SC 5, the standardization committee responsible for drafting the European Timber Design standard, has decided to establish a Working Group 7 "Reinforcement" on this item. This chapter presents the approach to this task, the work items, the work plan, the structure as well as the design approaches, and related background information for the proposed Eurocode 5 section.

1 Introduction

The reinforcement of timber structures has seen considerable research and development in recent years, as compiled in the RILEM and COST Action state-of-the-art reports [1]. Recent developments such as self-tapping screws and screwed-in threaded rods offer great potential in their use as reinforcement. For their use in construction works, it has to be verified that essential requirements like mechanical resistance, stability and safety are met. The required performance is commonly verified by complying with the corresponding harmonized technical rules for their structural design as well as for products used in the construction work. In cases where harmonized technical rules or technical approvals are not available, an approval in the individual case or comparable (depending on the particular nation's building regulations) has to be sought. Many reinforcement methods still lack harmonized technical

In memory of Alfons Brunauer (1963–2018)

P. Dietsch (✉)
Unit of Timber Engineering, Department of Structural Engineering and Material Sciences, University of Innsbruck, Innsbruck, Austria
e-mail: philipp.dietsch@uibk.ac.at

J. Branco et al. (eds.), *Reinforcement of Timber Elements in Existing Structures*,
RILEM State-of-the-Art Reports 33,
https://doi.org/10.1007/978-3-030-67794-7_6

rules, and most current design standards do not comprise approaches to design reinforcements for timber members. This is also the case for the current (2004) edition of the European timber design standards, Eurocode 5 [2]. The use of reinforcements is standardized only in a few European countries by means of non-contradictory information (NCI) given in the National Annexes (NAs) to Eurocode 5, most notably the German [3] and Austrian [4] NAs.

Closing the gap between recent developments and, on the one hand, practical needs and, on the other, missing standardization, reinforcement for stresses perpendicular to the grain was classified as high priority when defining the list of work items for the upcoming revision of Eurocode 5 [5]. In 2011, the European standardization committee responsible for Eurocode 5, CEN/TC 250/SC 5, decided to form a Working Group (WG) 7 "Reinforcement". In addition, reinforcement of timber members was prioritized for Phase 1 (of 4 phases) of the standardization work to be authorized by the European Commission. The contracts for this work were signed in 2014, enabling the formation of Project Teams (PT) that were mandated to draft specific Eurocode sections.

2 Approach to the Standardization Work

Standardization is the culmination of successful research and development that have seen positive application and acceptance in practice (see Fig. 1). According to the European position on future standardization [6], harmonized technical rules shall be prepared for "common design cases" and shall contain "only commonly accepted results of research and validated through sufficient practical experience". The target audience for such rules is "competent civil, structural and geotechnical engineers, typically qualified professionals able to work independently in relevant fields".

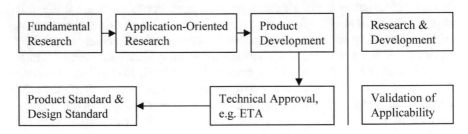

Fig. 1 Development of products or methods and their legalization

3   Organization of the Standardization Work

Different committees and groups of experts contribute to the European standardization in the field of the design of structures (see Fig. 2 for the example of timber structures). In this section, a short description of the main structure and organization of these committees and groups is given. For an in-depth description, the interested reader is referred to [7].

CEN/TC 250 is the head committee, responsible for the development and definition of the design rules of common building and civil engineering structures. This committee is substructured into 11 subcommittees (SCs), each SC being responsible for the development and revision of one Eurocode. CEN/TC 250/SC 5 is responsible for Eurocode 5 (EN 1995). The members of these SCs are delegates sent by National Standardization Bodies (NSBs) that are members of the CEN (Comité Européen de Normalisation/ European Committee for Standardization).

For the technical work, each SC is supported by WGs that deal with specific items. Within CEN/TC 250/SC 5, WG 7 is responsible for reinforcement of timber structures (see [7] for a full overview). The WGs are responsible for the development of the work programme, i.e. the items to be covered under their responsibilities. In this regard, the WGs are meant to serve as platforms for technical discussions that would result in technical proposals (methods, design approaches, design equations and details)

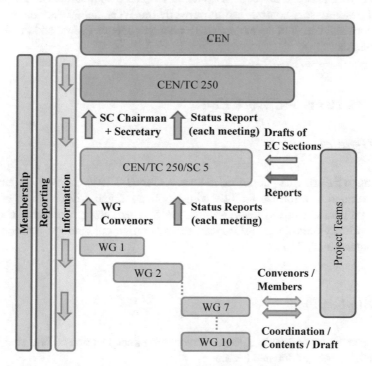

Fig. 2 Organization, responsibilities and reporting within the CEN

for the section(s) under their responsibilities. To achieve this objective, the National Standardization Bodies (NSB) send experts to the WGs. CEN/TC 250/SC 5/WG 7 "Reinforcement" currently has 20 members (experts and observers); about six experts contribute actively to the work.

The drafting of the standard text based on the technical proposals developed and agreed between the WGs is the responsibility of the Project Teams (PTs), each of which consists of five members and one leader. The work of the PTs is supported by the European Commission, hence it is established in a tender based on individual applications. Within a given time frame (in this case 42 months), the PTs have to deliver a draft of a new or revised Eurocode or a specific section of such a draft. In other words, the PTs have to bring the technical proposals into the standard text, including harmonized notations, terminologies and references, and adhere to the principles of "Ease-of-Use" [6]. In addition, the PTs have to develop "background documents" describing the technical reasoning and scientific background of all new or changed technical contents under their scope. Of the members of the PT, SC5.T1 "Cross-laminated timber and reinforcement", three members (A. Brunauer, T. Wiegand and the author) were actively involved in the drafting of the section on reinforcement, while four members (G. Schickhofer, R. Tomasi, T. Wiegand and the author) actively contributed to the drafting of the section on cross-laminated timber (CLT).

The liaison between the SCs, WGs and PTs can be summarized as follows: the SC is the supervisory control institution while the WGs and PTs are the executive institutions saddled with developing the technical contents (WGs) and the drafts of the standards (PTs).

4   Work Items and Work Plan

4.1   General

Adhering to the principles described in Sect. 2, CEN/TC 250/SC 5/WG 7 "Reinforcement" decided to prioritize the following application and reinforcement methods for the preparation of the revised Eurocode 5. These items were also classified as high priority during a pan-European survey carried out amongst a multitude of stakeholders in 2010 [5].

4.2   Applications

- Reinforcement of double tapered, curved and pitched cambered beams
- Reinforcement of notched beams
- Reinforcement of holes in beams

- Reinforcement of connections with a force component perpendicular to the grain
- Reinforcement of dowel-type connections ($n = n_{ef}$)
- Reinforcement of members with concentrated compression stresses perpendicular to the grain

Compared to the state of the art in research (see the RILEM state-of-the-art report or [1]), this list is less comprehensive. For example, reinforcement to increase bending or shear capacity and reinforcement of carpentry connections are not included. The reason is the requirement discussed in Sect. 2, which states that only rules for "common design cases" based on "commonly accepted results of research" that are "validated through sufficient practical experience" shall be included in the standards.

4.3 Materials

- Self-tapping screws (STS) or screwed-in threaded rods with wood screw thread
- Glued-in rods (GiR)
- Glued-on timber, plywood, and laminated veneer lumber (LVL)

Compared to the state of the art in research (see the RILEM state-of-the-art report or [1]), this list again is less comprehensive. For example, FRP and nanotechnology reinforcements are not included. The choice of materials is explained by the precondition that (1) test procedures and (2) a product standard or Technical Approvals for the product/material are available. Without these documents, rules in a design standard cannot be used since the basic input parameters would be missing. This situation can best be described by a 3-step pyramid (see Fig. 3).

This pyramid is based on (1) test standards (containing rules on how to test products). Relating to these, product standards (2) are developed (giving strength and stiffness parameters, boundary conditions, and rules for production and quality control). The design standards (3) represent the tip of this pyramid (providing design equations and formulating specific requirements, e.g. spacing, edge distance, minimum anchorage length, etc.). When developing design rules, it is a precondition to also develop (1) test procedures and (2) a product standard for the product or system used. Without the latter, rules in a design standard cannot be used since the basic

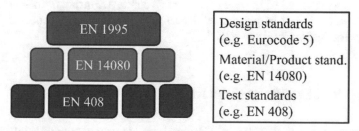

Fig. 3 Sketch of the 3-step pyramid applied in standardization for the construction sector [8]

parameters would be missing. In other words, the pyramid will not be complete since one element is missing. For further information on this topic, the interested reader is referred to [8], where two widely used reinforcement materials (STS and GiR) are discussed with reference to the currently existing standards and standards required to complete the pyramid.

For applications or materials for which the abovementioned requirements cannot be met but for which sufficient accepted results are available, the CEN offers the possibility of Technical Specifications (TS). A TS is a normative document whose development can be envisaged in anticipation of future harmonization and standardization, or for providing specifications for evolving technologies. A TS is established by a CEN TC, SC or WG, and approved through the same system of weighted votes as for the European standards ($\geq$55% of the votes cast and $\geq$65% of the population of the countries of the members having voted positively) by the CEN national members. The TS is then announced at the national level and may be adopted as a national standard. A TS must, however, not conflict with a European Standard, i.e. if a conflicting EN is published, the TS is withdrawn [9].

5 Work Plan

While most WGs are permanent technical bodies within the CEN, the time for the PTs to achieve their work plan is limited. Six months after the official start, a PT has to deliver the first draft. The corresponding CEN/TC 250/SC has two months to review the draft and make comments. The SC can call upon the responsible NSB for additional review and comments. The PT is requested to respond to all comments received while working on the second draft and to implement all comments and proposals deemed useful and technically sound. The second draft has to be delivered one year after the first and must pass through the same steps as the first. The third draft has to be delivered six months after the second. These will be directly forwarded to the NSB for a three-month enquiry. Following that, the PT have two months to prepare the final documents, taking into account the comments received from the NSB. The delivery of the final documents and the background documents, which marks the end of the work by the PT, is 42 months after the official beginning of the work by the PT. Project Team 1 delivered the final draft on "Reinforcement" in May 2018, which marked the end of the team's contract. Further refinements on the "Reinforcement" draft before consolidation with EN 1995-1-1 (expected summer 2021) are the responsibility of WG 7.

6 Proposed Standard Structure

The current version of EN 1995-1-1 [2] does not contain provisions on reinforcement. Hence, a decision on the structure of this new section had to be taken. The obvious approach would have been to write a separate but continuous section on the design of reinforcement for timber members. This solution, however, would not have fully met the designers needs in terms of applicability and navigation, i.e. it would not have fully agreed with the principles of "Ease-of-Use". The proposal, which has been accepted by CEN/TC 250/SC 5, was to integrate the provisions on reinforcement into the existing main part, i.e. following the sequence of a typical design task: general considerations—design of members in the unreinforced state—design of reinforcement for these members (see Fig. 4).

7 Contents of the Draft "Reinforcement"

7.1 General

In what follows, the core contents of the sections on reinforcement are given in italics, followed by relevant background information on these clauses. For a comprehensive overview of the current state of the art in the design of reinforcement, including design equations and extensive background information, the interested reader is referred to [10] and [11]. The figures shown do not represent the figures for the standard text as they also include graphical representations used to exemplify background information. Since the strength verifications required for the reinforcement are rather independent of the member or detail to be reinforced, these are presented in a consolidated form in this first section (7.1 "General"). An exception is the reinforcement of members with compression stresses perpendicular to the grain, hence (and different to the proposed standard structure) the proposed standard text and necessary amendments to existing sections will be presented in a consolidated form at the end of this section.

Standard text (Sect. 8.4.1):

- *In the following clauses, the tensile capacity perpendicular to the grain of the timber is not taken into account for the determination of the load on the reinforcement.*
- *Pitched cambered beams should be reinforced for tensile stresses perpendicular to the grain. Where the design tensile stresses perpendicular to the grain exceed 60% of the design tensile strength perpendicular to the grain of curved and double tapered beams, these should be reinforced.*
- *NOTE: Reinforcement of notches and holes in beams leads to more robust long-term performance of members, especially in the case of large member sizes and/or large expected changes in the timber moisture content.*
- *Notched members, holes in beams, double tapered beams, curved beams and pitched cambered beams assigned to Service Class 3 should be reinforced.*

Background of the clauses given above:

Section	Title
8	Ultimate limit states
8.1	Design of cross-sections subjected to stress in one principal direction
…	
8.1.5	Compression perpendicular to the grain
8.1.5.1	General
8.1.5.2	***Reinforcement of members with compression stresses perpendicular to the grain***
…	
8.4	**Design of members with special requirements based on geometrical form**
8.4.1	***General***
8.4.2	***Effects of moisture content changes on reinforced members***
8.4.3	Design of cross-sections in members with varying cross-sections or curved shapes
8.4.3.1	General
8.4.3.2	Single tapered beams
8.4.3.3	Double tapered, curved and pitched cambered beams
8.4.3.4	***Reinforcement of double tapered, curved and pitched cambered beams***
8.4.4	Notched members
8.4.4.1	General
8.4.4.2	Beam with a notch at the support
8.4.4.3	***Reinforcement of notches***
8.4.5	Holes in beams
8.4.5.1	General
8.4.5.2	Holes in beams with rectangular cross-sections
8.4.5.3	***Reinforcement of holes in beams***
…	
10	Joints and Connections
10.1	General
10.1.4	Connection force at an angle to the grain
10.1.4.1	Design of a connection with a tensile force component perpendicular to the grain
10.1.4.2	***Reinforcement of a connection with a tensile force component perpendicular to the grain***
…	
10.5	Bolted connections
10.5.1	Laterally loaded bolts
10.5.1.1	***General and bolted timber-to-timber connections***

Fig. 4 Proposed structure for sections on reinforcement (new sections in **bold**, sections on reinforcement in ***bold italic***)

For the approaches given, the tensile capacity perpendicular to the grain of the timber is neglected, i.e. a cracked cross-section is assumed in the direction of the tensile stresses perpendicular to the grain. This is different to the method presented in [12], in which only the force components exceeding the tensile strength perpendicular to the grain of the timber are applied for the design of the reinforcement. Before cracking of the cross-section perpendicular to the grain, a proportional share of the tensile stresses perpendicular to the grain is transferred by the timber. The share depends on the stiffness of the reinforcement embedded in or around the timber compared to the stiffness of the timber member, and also on the distance of the cross-section under consideration to the next reinforcing element.

Even if the verification of systematic, load-dependent tensile stresses perpendicular to the grain can be met, it is state of the art to reinforce double tapered, curved and pitched cambered beams against tensile stresses perpendicular to the grain. The reason is the superposition of the load-dependent stresses on moisture-induced stresses perpendicular to the grain due to, for example, changing climatic conditions or a drying of the beam after the opening of the building (see [13] for example). In the absence of a method to reliably predict the magnitude of tensile stresses perpendicular to the grain, it was customary to apply reinforcement if the maximum load-dependent tensile stresses perpendicular to the grain exceeded 75% of the permissible tensile strength perpendicular to the grain. With the transition from the system of permissible stresses to the semi-probabilistic system, this approach was transferred into the requirement that in unreinforced beams, the maximum load-dependent design tensile stresses perpendicular to the grain should not exceed 60% of the design tensile strength perpendicular to the grain.

Since end-grain is exposed bare at a notch and in holes, the superposition of moisture-induced stresses on load-dependent tensile stresses perpendicular to the grain around notches and holes can be significant [14]. Therefore, many authors recommend that notches and holes in beams should always be reinforced.

Members in SC 3 should always be reinforced. The reason are the strong moisture changes in SC 3, leading to moisture-induced stresses in the timber member of magnitudes that leave no or only marginal capacity for systematic, load-dependent tensile stresses perpendicular to the grain. For members in SC 3 such as curved and pitched cambered beams, it is recommended to apply external plane reinforcement glued onto the entire surface area under tensile stresses perpendicular to the grain. It should be attempted to enable a classification of the member in SC 2 by e.g. constructive protection measures.

- *The following internal, dowel-type reinforcement may be applied:*

  – *fully threaded screws in accordance with EN 14592 or the European Technical Assessment;*
  – *screwed-in threaded rods with wood screw thread in accordance with the European Technical Assessment;*
  – *glued-in threaded or ribbed steel rods.*

- *The following plane reinforcement may be applied:*

  – *glued-on plywood or solid wood panels in accordance with EN 13986;*
  – *glued-on structural laminated veneer lumber in accordance with EN 14374;*
  – *glued-on laminations made from structural solid timber in accordance with EN 14081-1, plywood in accordance with EN 13986 or structural laminated veneer lumber in accordance with EN 14374;*

- *pressed-in punched metal plate fasteners.*

- *The reinforcement shall be applicable for the timber product and the Service Class of the reinforced timber element.*

The list of applicable internal or external reinforcements is, amongst other factors, based on the necessity for a continuous interconnection between the timber and the reinforcement as well as for sufficient stiffness of this connection (to prevent cracking). Due to the latter, perforated metal plates or wood-based panels, nailed onto the timber member, are not adequate reinforcements (see [14] and [15] for example).

- *The distance between the peak tensile stresses perpendicular to the grain and the dowel-type reinforcement should be minimized but not less than the following minimum values.*
- *The spacing between glued-in threaded rods, a_2, should not be less than $3 \cdot d$. The edge distance in the grain direction, $a_{3,c}$, as well as the edge distance perpendicular to the grain, $a_{4,c}$, should not be less than $2,5 \cdot d$.*
- *For fully-threaded screws and screwed-in threaded rods, the spacing rules should be taken from Table 8.6 in [2] or from the European Technical Assessment.*
- *For inclined dowel-type reinforcement, the spacing may be determined based on the centre of gravity of the dowel-type reinforcement in the section of the timber member (see Figs. 6 and 7).*
- *The reduction in the cross-sectional area due to internal reinforcement should be considered in the design of the timber member.*
- *In block glued members (see [16] for example), each component within the block should be reinforced either by internal dowel-type reinforcement or by plane reinforcement glued to both side faces of each component. The reduction in the cross-sectional area due to glued in plane reinforcement should be considered in the design of the block glued member.*

The reinforcing effect of dowel-type or plane reinforcement is strongly dependent on the distance between the reinforcement and the location of peak stresses (for tension perpendicular to the grain or shear for example). Edge and end distances of glued-in steel rods are partly reduced compared to the minimum edge and end distances given in Chapter 8 of EN 1995-1-1:2004 [2], since such reinforcements are loaded by axial forces and their continuous interconnection with the wood prevents splitting [15]. Inclined dowel-type reinforcement makes it possible to meet the requirement of reducing as much as possible the distance between the reinforcement and the location of peak stresses, hence the possibility of applying inclined dowel-type reinforcement with distance requirements based on the position of the centre of gravity of the dowel-type reinforcement in the timber member under consideration, given in [2], was introduced. The reinforcing effect of the applicable reinforcement elements on the width of a timber member is limited, hence each component of a block-glued timber member should be reinforced separately.

- *The design tensile force in a reinforcement should satisfy Formula (1):*

$$\frac{F_{t,90,Ed}}{F_{t,90,Rd}} \leq 1,0 \tag{1}$$

where

$F_{t,90,Ed}$ *is the design tensile force in the reinforcement, according to the formulae given in* Sects. 7.3–7.7;

$F_{t,90,Rd}$ *is the design tensile resistance of dowel-type or plane reinforcement, according to Formulae (2)–(4)*

The design resistance of dowel-type or plane reinforcement should be taken as the minimum value obtained from Formulae (2)–(4):

– *For fully threaded screws or fully threaded rods with wood screw thread* (see also Sect. 8.7 in [2]):

$$F_{t,90,Rd} = n_r \cdot \min \begin{cases} f_{ax,d} \cdot d \cdot l_{ad} \\ f_{tens,d} \end{cases} \tag{2}$$

– *For glued-in steel rods:*

$$F_{t,90,Rd} = n_r \cdot \min \begin{cases} f_{b1,d} \cdot \pi \cdot d \cdot \ell_{ad} \\ f_{yb,d} \cdot A \\ 0,9 \cdot f_{ub,d} \cdot A_S \end{cases} \tag{3}$$

– *For glued-on plane reinforcement:*

$$F_{t,90,Rd} = n_r \cdot \min \begin{cases} f_{b2,d} \cdot \ell_{ad} \cdot b_r \\ \frac{f_{t,d}}{k_k} \cdot b_r \cdot t_r \end{cases} \tag{4}$$

with:

$$l_{ad} = \min \begin{cases} l_{ad,t} \\ l_{ad,c} \end{cases} \tag{5}$$

(determined in accordance with the geometry of the detail

to be reinforced, see Fig. 5–Fig. 9 for example)

where

n_r *is the number of reinforcing elements* (typically 2, resp. 4 with the exception of curved and pitched cambered beams, where $n = 1$);

$f_{ax,d}$ *is the design withdrawal strength of the fully threaded screw/rod;*

$f_{tens,d}$ *is the design tensile capacity of the fully threaded screw/rod;*

$f_{ub,d}$ *is the design ultimate strength of the steel rod;*

$f_{yb,d}$ *is the design yield strength of the steel rod;*

$f_{b1,d}; f_{b2,d}$ *is the design strength of the glue line;*
$f_{t,d}$ *is the design tensile strength of the plane reinforcement;*
d *is the outer thread diameter of the fully threaded screw or steel rod*
            (≤ 20 mm);
A *is the gross cross-section of the steel rod (see [17]);*
A_S *is the tensile stress area of the steel rod (see [17]);*
ℓ_{ad} *is the relevant effective anchorage length, glued-in length, relevant*
            depth of plane reinforcement;
$\ell_{ad,t/c}$ *is the relevant effective anchorage length above (below) the axis*
            prone to, splitting;
b_r *is the width of the plane reinforcement;*
t_r *is the thickness of the plane reinforcement;*
k_k *is a factor to account for non-uniform distribution of stresses in*
            the plane reinforcement. Without further verification, $k_k = 2,0$ may
            be assumed (for reinforcement of connections with a tensile force
            component perpendicular to the grain, $k_k = 1,5$ may be assumed; for
            reinforcement of curved or pitched cambered beams, $k_k = 1,0$ may be
            assumed);

- *Reinforcement with punched metal plate fasteners should be designed in analogy
  to Formula (4), using the anchorage strength, and should be placed according to
  the rules for plane reinforcement given in the following sections.*

The assembly of all equations to determine the resistance of the reinforcement in
one place was undertaken with the aim of:

- realizing a homogeneous set of equations independent of the member or detail to
  be reinforced,
- enabling a better overview of equations to determine resistances that include all
  corresponding factors (ease-of-use),
- reducing the length of the document by 25% compared to the preceding NAs.

The verifications follow the verification procedures introduced in EN 1993-1-
8:2005 [17] (for steel rods), Chapter 8 of EN 1995-1-1:2004 [2] (for fully threaded
screws and rods with wood screw thread) and Chapter 6 of EC 5 [2] (for plane
reinforcement). The applicable anchorage length is the shorter of the two lengths
above respectively below the location of crack onset (which is assumed to be identical
to the location of peak stresses). The factor k_k is applied to take into account the
characteristics of the non-uniform distribution of stresses and the concentration of
stresses at the panel edge facing the peak stresses in the timber member [18]. The
effective number of fasteners, n_{ef}, does not apply. The reason is that the load transfer
mechanism is different compared to connections with axially loaded screws or rods
("passive" vs. "active" application). In the majority of cases, only 1 or 2 screws or
rods are used for reinforcement of notched members and holes in beams. In curved
and pitched cambered beams, each reinforcing element transfers the released stresses
in a specific area, hence the load transfer mechanism is different to that for a group
of screws sharing one load in a connection.

7.2 Effects of Changes in the Moisture Content

Standard text (Sect. 8.4.2):

- *The effects of changes in the moisture content of the timber (e.g. shrinkage cracks) shall be taken into account.*

Background of the clause given above:

Changes in the moisture content of wood lead to changes in virtually all physical and mechanical properties (e.g. strength and stiffness properties) of the wood. Additional effects of changes in the wood moisture content are the shrinkage or swelling of the material and the associated internal stresses. If these stresses locally exceed the very low tensile strength perpendicular to the grain of the wood, the result will be a stress relief in the form of cracks, which can reduce the load-carrying capacity of the structural timber elements in, for example, shear or tension perpendicular to the grain. Multiple evaluations of damage in timber structures (see [19–21]) show that a prevalent type of damage is pronounced cracking in timber elements. Almost half of the damage in large-span glued-laminated timber structures can be attributed to a low or high moisture content or severe changes of the same.

- *The effects of changes in the moisture content of timber should be minimized. Potential measures to reduce the effects of changes in the moisture content include:*

  - *Before being used in construction, the timber should be dried as near as practicable to the moisture content appropriate to its climatic condition in the building in use, unless the structure is able to dry without significant effects on the load-carrying capacity of its members;*
  - *During transport, storage and assembly, the timber should be protected to minimize detrimental changes in its moisture content;*
  - *In dry environments, controlled drying of the timber to service conditions should be planned.*

- *For structures or members sensitive to moisture changes, temporary moisture control is recommended, until the expected equilibrium moisture content is reached.*

Effects of changes in the moisture content include changes in strength and stiffness properties (covered by k_{mod}). Another effect of changes in the moisture content are shrinkage cracks. Shrinkage cracks can be attributed to two different phenomena.

1. Large moisture gradient over the timber cross-section due to strong and fast wetting or drying of the timber member (the latter prevailing in closed and heated buildings), occurring, for example, throughout the process production—transport—storage—assembly—interior works—opening—operation (heating). Careful planning and moisture control during this process are recommended, especially if a dry environment is to be expected for the finished

building. Specifications on moisture control could be given in an execution standard for timber structures.

2. Prevention of free shrinkage or swelling deformation of the cross-section by restraining forces from, for instance, connections covering larger heights or dowel-type reinforcements. In these cases, equilibrium of tensile and compressive moisture induced stresses is impeded, resulting in stresses of higher magnitude and, eventually, in deep shrinkage cracks.

Being that there is currently no method to reliably predict the magnitude of tensile stresses perpendicular to the grain due to moisture changes, it was decided to introduce the term *effects of changes in the moisture content*.

- *The effects of reinforcement (or connections), which restrain moisture-induced deformations of the timber member, should be minimized.*
- *Potential measures to reduce restraining effects of reinforcement include:*
  - *increasing the spacing between the reinforcements;*
  - *reducing of the height of the reinforced area of the timber member;*
  - *reducing the angle between the dowel-type reinforcement and the grain direction of the timber member.*
- *Where reinforcement is necessary in applications in permanently dry or frequently changing climate, external plane reinforcement glued onto the entire surface area under tensile stresses perpendicular to the grain should be preferred, as it decelerates the process of moisture changes or drying of the timber member.*
- *NOTE: Where external plane reinforcement is used to cover peak tensile stresses perpendicular to the grain in vicinity of the end grain (e.g. notches and holes in beams), adequate surface treatment of the end grain leads to a reduction of the restraining effect of the plane reinforcement especially in permanently dry climates.*

The restraining effect of dowel-type reinforcement was experimentally and analytically investigated in [22] and [23], and the studies demonstrated the positive effect of measures such as increased distance, reduced height and inclined positioning of dowel-type reinforcement. Attention should be paid to the additional stresses induced by the inclination of the reinforcement in a deformed timber beam (positive compression stresses perpendicular to the grain in the case of decreasing inclination, i.e. angle between load and grain, in the deformed shape vs. detrimental tensile stresses perpendicular to the grain in the case of increasing inclination in the deformed shape) (see [24]).

In the original draft, the proposed clauses on the effects of changes in the moisture content are separated into general clauses applicable to all timber elements (proposed as a new section, *2.3.3 Effects of moisture content changes*, under Section *2.3 Basic Variables*) and clauses applicable to reinforced timber elements (proposed for Section *6.4.2 Effects of moisture content changes on reinforced beams*). For reasons of representation, this differentiation has been omitted in this contribution.

7.3 Reinforcement of Double Tapered, Curved and Pitched Cambered Beams

Standard text (Sect. 8.4.3.4):

- *For beams on which reinforcement to carry the full tensile stresses perpendicular to the grain is applied, the design tensile force in the reinforcement, $F_{t,90,Ed}$, should be calculated as follows:*

$$F_{t,90,Ed} = k_{ka} \cdot \sigma_{t,90,d} \cdot b \cdot a_1 \tag{6}$$

where

$\sigma_{t,90,d}$ *is the design tensile stress perpendicular to the grain (acc. to Eq. 6.54 in [2]);*

b *is the beam width;*

a_1 *is the spacing of the reinforcement along the longitudinal direction of the beam at the height of its axis (see Fig. 5);*

k_{ka} *is a factor to account for the distribution of tensile stresses perpendicular to the grain along the beam axis*
$k_{ka}= 1,0$ for curved beams and for the inner quarters of the length of the volume exposed to tensile stresses perpendicular to the grain, measured from the apex in double tapered and pitched cambered beams;
$k_{ka}= 0,67$ for the outer quarters of the area exposed to tensile stresses perpendicular to the grain, measured from the apex in double tapered and pitched cambered beams;

Background of the clause given above:

The approach given is based on the integration of the sum of tensile stresses perpendicular to the grain in the plane of zero longitudinal stresses. Since, in most design standards (e.g. [2]), only the formulae to determine the maximum tensile stresses perpendicular to the grain in the apex are given, the distribution of tensile stresses perpendicular to the grain along the beam axis has to be accounted for in

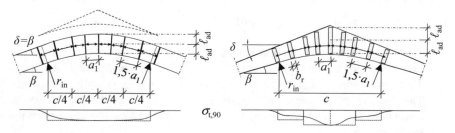

Fig. 5 Curved and pitched cambered beams: stress distribution, reinforcement and geometries. Curved beam with an indication for the mechanically jointed apex

a simplified format. Depending on the form and loading of the beam, the tensile stresses perpendicular to the grain decrease with increasing distance from the apex (an exception being curved beams with mechanically jointed apexes, i.e. secondary apexes, see subsequently). For simplification, the full tensile stresses perpendicular to the grain are used to design the reinforcement in curved beams and the inner quarters of the area exposed to tensile stresses perpendicular to the grain are used for the same purpose in double tapered and pitched cambered beams. In the outer quarters of double tapered and pitched cambered beams, the tensile stresses perpendicular to the grain are assumed to reach 2/3 of the maximum tensile stresses perpendicular to the grain (see Fig. 5).

- *For curved or pitched cambered beams with a mechanically jointed apex, reinforcement should be designed for:*

  - *tensile stresses perpendicular to the grain at the inflection points (secondary apex at the end of the mechanically jointed apex), and*
  - *tensile stresses perpendicular to the grain from the curvature in the apex.*

  The reinforcement at the inflection points should cover a length of at least $2 \cdot h_{ap}$ in the direction of the apex and $1 \cdot h_{ap}$ in the direction of the beam end. The reinforcement from the curvature in the apex should be arranged in the remaining curved parts. Between both areas, the spacing between the reinforcement may be linearly graded. If the tensile stresses perpendicular to the grain from the curvature in the apex are higher than the tensile stresses perpendicular to the grain at the inflection points, the associated reinforcement should be arranged over the whole curved length.

Curved beams with a mechanically jointed apex (see Fig. 5) are neither regulated in EN 1995-1-1 [2] nor in NCI, such as [3] and [4]. Nevertheless, these beams represent the most widely utilized form in practice. The most common type are curved beams with a raised dry joint and a mechanically jointed apex to realize the form of a pitched cambered beam. The top edge of the beam features a shorter curved length compared to the curved length of the bottom edge, leading to the so-called secondary apexes at the transition points between the curved upper edge and the straight upper edge of the beam. The approach to the design of curved beams with a mechanically jointed apex and their reinforcement is to examine and verify two different cross-sections:

1. The apex formed by the curved part of the beam and to be designed in analogy to curved beams;
2. The secondary apex located at the transition point between the curved upper flange and the straight upper flange at the end of the mechanically jointed apex. This cross-section is to be designed in analogy to pitched cambered beams. The distance of the secondary apexes is limited to a minimum of twice the beam height in the apex to prevent the superposition of peak stresses from the secondary apex.

The length of the area to be covered by the reinforcement determined for the two cross-sections is dependent on the length of the area of decreasing stresses. The design approach is in the style of the information and results presented in [25] and [26]. The distance requirements implicitly given by the factor k_{ka} do not apply for curved beams with a mechanically jointed apex.

- *The spacing of the reinforcement parallel to the grain, a_1, may be adapted according to the distribution of tensile stresses perpendicular to the grain along the length of the volume under the tensile stresses perpendicular to the grain.*
- *Internal, dowel-type reinforcement should cover the full height of the beam excluding the outer laminations in bending tension. One reinforcing element should be placed in the cross-section below the apex respectively secondary apex (inflection point). The spacing, a_1, at the top side of the beam should not be less than 250 mm but not greater than $0,75 \cdot h_{ap}$.*
- *Plane reinforcement, e.g. panels or laminations, should be glued to both sides of the member and should cover the full height of the beam; at maximum it should exclude the outer laminations (see Fig. 5).*

The requirement to place the reinforcement as close as possible to peak stresses also applies here. The spacing between the reinforcement is limited to ensure that the reinforcing effect is assured over the whole beam length exposed to tensile stresses perpendicular to the grain (see Fig. 5). The timber cross-section should not be reduced by reinforcement in the vicinity of maximum tensile bending stresses (i.e. the outermost laminations). The length requirements are meant to exclude failure in tension perpendicular to the grain in the beam below or above the reinforcement.

7.4 Reinforcement of Notches

Standard text (Sect. 8.4.4.3):

- *The following rules apply for reinforced notches in members with rectangular cross-section from kiln-dried solid timber, glulam and laminated veneer lumber. For members with a rectangular notch on the same side as the support and a/h ≤ 0,4 (see Fig. 6), the reinforcement may be designed for a design tensile force $F_{t,90,Ed}$:*

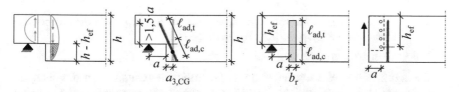

Fig. 6 Notched beams: distribution of shear stresses (left), dowel-type reinforcement (middle left), plane reinforcement (middle right) and reinforced connector/group of fasteners (right)

$$F_{t,90,Ed} = 1,3 \cdot V_d \cdot \left[3 \cdot (1 - \alpha)^2 - 2 \cdot (1 - \alpha)^3\right] \qquad (7)$$

where
$\alpha = h_{ef}/h$ *(see Fig. 6)*

Background of the clause given above:

The tensile force perpendicular to the grain, $F_{t,90,Ed}$, can be approximated by the integration of the shear stresses below the notch, between the loaded edges and at the corner of the notch (see Fig. 6). A more detailed analysis of the magnitude of the tensile stresses perpendicular to the grain around the notch has shown that these stresses are even higher [27]. For the relationships $a \leq h_{ef}/3$ and $h_{ef}/h \geq 0,5$, the tensile force perpendicular to the grain, $F_{t,90,Ed}$, can be sufficiently estimated by applying an increase factor of 1,3. The reinforcement of notches can be transferred to connections, resulting in comparable behaviour (e.g. dovetail connections or modern primary beam—secondary beam connectors).

- *The reinforcement should cover the full height of the notched edge $\ell_{ad,c} = (h - h_{ef})$. The minimum length $\ell_{ad,t} = min\{\ell_{ad,c}; 1,5 \cdot a\}$ (see Fig. 6).*
- *Where the tensile force, $F_{t,90,Ed}$, according to Formula (7), is carried by internal dowel-type reinforcement, only one row of the internal reinforcing elements at a distance $a_{3,c}$ from the edge of the notch should be considered (see Fig. 6). The dowel-type reinforcement may be inclined to reduce the distance between the peak tensile stresses perpendicular to the grain and the dowel-type reinforcement (see Fig. 6).*
- *Where the tensile force, $F_{t,90,Ed}$, according to Formula (7), is carried by internal dowel-type reinforcement oriented perpendicular to the grain, the load-carrying capacity is limited to twice the load-carrying capacity of the unreinforced notched beam. In addition, the shear stresses (Formula (6.13) in [2]) should be satisfied in the notched part.*
- *Plane reinforcement, e.g. panels or laminations, should be glued to both sides of the member, with the following limits:*

$$0,25 \leq \frac{b_r}{h - h_{ef}} \leq 0,5 \qquad (8)$$

where

b_r *is the width of the reinforcement panel or lamination in the direction of the beam axis at the side of the notch;*
h and h_{ef} *see Fig. 6.*

The depth of the reinforcement should be sufficient enough to ensure adequate load transfer into the reinforcement and from the reinforcement into the support. The required length for the latter requirement is based on a load-distribution angle of 45°. Failure in tension perpendicular to the grain below the reinforcement should be prevented.

Only one row of dowel-type reinforcement at a distance $a_{3,c}$ should be considered as reinforcement. The distance between the dowel-type reinforcement and the notch, $a_{3,c}$, should be as small as possible. The reason is the limited distribution length of the tensile stresses perpendicular to the grain outside the corner of the notch. This can be achieved by inserting the screws at an inclined angle, as the distance requirements are based on the position of the centre of gravity of the dowel-type reinforcement in the timber member under consideration ($a_{3,CG}$, see Fig. 6). Rounding of the corners of the notch leads to a reduction in peak stresses.

The limitation of the load-carrying capacity of notched members reinforced with dowel-type reinforcement arranged perpendicular to the grain is based on [28], where it was experimentally and analytically verified that the load-carrying capacity of reinforced notched members is not infinite but limited by the magnitude of the shear component on fracture of the notched members.

The applicable width of reinforcement panels is limited due to the limited distribution length of the tensile stresses perpendicular to the grain outside the corner of the notch. In addition, this limitation is also implicitly directed at assuring panels of adequate thickness to prevent failure due to stress singularities at the notch. To facilitate screw-press-gluing and to realize robust plane reinforcement, a minimum panel width is recommended. Irrespective of this recommendation, the abovementioned upper limits of panel width should to be applied in stress verifications.

7.5 Reinforcement of Holes in Beams

Standard text (Sect. 8.4.5.3):

- *The following rules apply for members, from kiln-dried solid timber, glulam and laminated veneer lumber, with rectangular cross-sections and reinforced holes which comply with the geometrical boundary conditions given in Table 1.*

Background of the clause given above:

The requirement to apply reinforced holes only in members from timber products that have been technically dried is based on the necessity to reduce moisture-induced deformations (shrinkage, including potential cracking) of reinforced timber members. The limitations of the permissible relative dimensions of the holes,

Table 1 Minimum and maximum dimensions of reinforced holes in beams with rectangular cross-sections

$\ell_v \geq h$	$\ell_z \geq h$, not less than 300 mm[c]	$\ell_A \geq h/2$	$h_{rl(ru)} \geq 0,25 \cdot h$	$a \leq h$ $a/h_d \leq 2,5$	$h_d \leq 0,3 \cdot h^a$
					$h_d \leq 0,4 \cdot h^b$

[a]*for internal dowel-type reinforcement*
[b]*for plane reinforcement, e.g. panels or laminations*
[c]*l_z is the clear spacing between two holes*

depending on the type of reinforcement, are described in [29] and [30]. The reason
for the restriction in hole height is, among others, the concentrated stresses in the
grain direction close to the hole edges above and below the hole due to the necessary
transfer of bending stresses around the hole. According to the results of experimental
and numerical investigations [31], these stresses can be several times higher than the
bending stresses at the beam edge.

- *The reinforcement of holes in beams should be designed for a tensile force perpendicular to the grain, $F_{t,90,Ed}$, which is composed of $F_{t,V,d}$ from the transfer of shear stresses and $F_{t,M,d}$ from the transfer of bending stresses. In the case of rectangular holes, the tensile force, $F_{t,90,Ed}$, should be assumed to act on planes, defined by the top and bottom faces of the hole, in the corners prone to tensile stresses perpendicular to the grain (see Fig. 7). In the case of circular holes, the tensile force, $F_{t,90,Ed}$, should be assumed to act under 45° from the centre of the hole with regard to the beam axis (see Fig. 7). All areas prone to splitting from tensile stresses perpendicular to the grain should be analyzed.*

$$F_{t,90,Ed} = F_{t,V,Ed} + F_{t,M,Ed} = \frac{V_d \cdot h_d}{4 \cdot h} \cdot \left[3 - \frac{h_d^2}{h^2}\right] + 0,09 \cdot \frac{M_d}{h} \cdot \left(\frac{h_d}{h}\right)^2 \quad (9)$$

where

h_r *min {h_{rl}; h_{ru}};*
h_d *is the hole depth;*

For the definitions of h, h_d, h_{rl}, h_{ru}, see Fig. 7.

The tensile force perpendicular to the grain, $F_{t,V,Ed}$, can be approximated by integration of the shear stress between the axis of the member and the expected location of the crack at the corner of the hole prone to cracking. The location of crack onset for round holes has been determined by numerical investigations [29] and has also been observed in tests. The tensile force perpendicular to the grain, $F_{t,M,Ed}$, has been derived from tests [32] but is still under investigation [29].

Note to the reader: The standard clauses for unreinforced holes in EN 1995-1-1, including the formulae to determine the tensile forces perpendicular to the grain at the

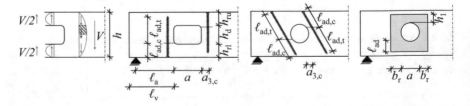

Fig. 7 Holes in beams: distribution of shear stresses (left), dowel-type reinforcement (middle left), inclined reinforcement (middle right) and plane reinforcement (right)

location of stress peaks, $F_{t,90}$, are currently under development in CEN/TC 250/SC 5. It is expected that the formulae will be adapted to recent research results. This includes an adjustment of the tensile force perpendicular to the grain from the transfer of bending stresses, $F_{t,M,Ed}$ [30], as well as the formulae for eccentrically arranged holes and groups of holes [31]. Due to the fact that this process was ongoing during the drafting of this chapter, it was decided to include the current state of the art in design formulae (see Formula 9). The applicability of the clauses given below is independent of the formulae to determine the tensile forces perpendicular to the grain, $F_{t,90,Ed}$.

- *Where the tensile force, $F_{t,90,Ed}$, is carried by internal dowel-type reinforcement, the relevant effective anchorage length ℓ_{ad}, should be taken as follows:*

  $$\ell_{ad} = h_{rl} \text{ or } h_{ru} \qquad\qquad \text{for rectangular holes;}$$
  $$\ell_{ad} = h_{rl} + 0,15 \cdot h_d \text{ or } h_{ru} + 0,15 \cdot h_d \text{ for circular holes;}$$

- *Where the tensile force, $F_{t,90,Ed}$, is carried by internal dowel-type reinforcement, only one row of the internal dowel-type reinforcing elements at a distance $a_{3,c}$ from the edge of the hole should be considered (see Fig. 7). The distance between the location of the maximum tensile stresses perpendicular to the grain and the dowel-type reinforcement should be minimized.*
- *The application of internal dowel-type reinforcement, positioned perpendicular to the grain, should be limited to locations in the timber member that are subjected to low shear stresses.*
- *In members with holes and internal dowel-type reinforcement, the increased shear stresses in the area of the edges of the holes should be accounted for. The maximum shear stress, τ_{max}, should be calculated as follows:*

$$\tau_{max} = \kappa_{max} \cdot \frac{1,5 \cdot V_d}{b_{ef} \cdot (h - h_d)} \tag{10}$$

where

$$\kappa_{max} = 1,85 \cdot \left[1 + \frac{a}{h}\right] \cdot \left(\frac{h_d}{h}\right)^{0,2} \tag{11}$$

b_{ef} *is the effective width (see 6.1.7. [2]), taking into account the impact of shrinkage cracks on shear capacity;*

For the definitions of a, h, h_d, see Fig. 7.
In the case of circular holes h_d may be replaced by $0,7 \cdot h_d$.
- *Where the shear verification with τ_{max} from Formula (10) is not fulfilled, internal reinforcement positioned perpendicular to the grain should not be used.*
- *For members from laminated veneer lumber (LVL) with holes it is recommended to use plane reinforcement, e.g. LVL with cross-veneers.*

- *Where internal dowel-type reinforcement is arranged according to Fig. 7, the spacing requirements given in Sect. 7.1 apply.*

The reinforcing effect of dowel-type or plane reinforcement is strongly dependent on the distance between the reinforcement and the location of peak stresses (in tension perpendicular to the grain or shear). To reduce this distance, the dowel-type reinforcement can be rotated to, for example, 60°. This arrangement has additional advantages, such as enabling the transfer of shear stresses and reducing the restraining effect in case of shrinkage (see Sect. 7.2).

The limited applicability of dowel-type reinforcement, arranged perpendicular to the grain, to areas exposed to low shear stresses is based on the fact that this arrangement restrains free shrinkage. This results in a reduced shear capacity of the reinforced timber member which, in the vicinity of holes, is exposed to increased shear stresses. Shear can only, to a small extent, be transferred by dowel-type reinforcement arranged perpendicular to the grain.

In the case of rectangular holes, it is necessary to take into account the increased shear stresses around the edges of the holes. A description as well as an associated design equation is given in [33]. In [15], it is recommended to apply the same verification for circular holes as well, although this yields results on the safe side. The same publication describes a method to verify the bending stresses above respectively below rectangular holes, including the additional longitudinal stresses from the frame action (lever of the shear force) around the hole (see also [32]).

- *Where the tensile force, $F_{t,90,Ed}$, is carried by plane reinforcement, the relevant effective anchorage length ℓ_{ad}, should be taken as follows:*

$$\ell_{ad} = h_1 \qquad \text{for rectangular holes;}$$
$$\ell_{ad} = h_1 + 0,15 \cdot h_d \quad \text{for circular holes;}$$

where

h_1 *is the depth of the plane reinforcement above or below a hole (see Fig. 7).*

- *The plane reinforcement, e.g. panels or laminations, should be glued to both sides of the member (see Fig. 7), with the following limits:*

$$0,25 \cdot a \leq b_r \leq 0,6 \cdot l_{t,90} \qquad (12)$$

where

$$l_{t,90} = 0,5 \cdot (h_d + h) \qquad (13)$$

and

$$h_1 \geq \max \begin{cases} 80 \text{ mm} \\ 0,25 \cdot a \end{cases} \qquad (14)$$

b_r *is the width of the plane reinforcement;*

$\ell_{t,90}$ *is the length under tensile stresses perpendicular to the grain;*

For the definitions of a, h_d, and h (see Fig. 7).

Depending on the type of product and type of application (e.g. screw press gluing), dimensions b_r (h_1) that exceed the upper limits given in Formula (12) (14) may be required. The width b_r applied in Formula (4) should not exceed the upper limit given in Formula (12).

- *For members from laminated veneer lumber (LVL) with holes, it is recommended to use plane reinforcement, e.g. LVL with cross-veneers.*

The specifications regarding the edge distance and the permissible number of rows of dowel-type reinforcement and the applicable width and the recommended minimum width of the reinforcement are based on the same conditions as described in Sect. 7.4.

Tests have shown that dowel-type reinforcement of holes in LVL beams do not necessarily increase the load-bearing capacity of the beams around the holes. It is therefore recommended to use LVL with cross veneers.

The requirement to produce rounded corners (radii $r/h_d > 0,1$; see Fig. 7) has been derived from the necessity for the dispersion of shear stresses at the corners in order not to exceed the shear stresses determined by Formula (11) together with practical considerations agreed by members of CEN/TC 250/SC 5/WG 7 "Reinforcement".

7.6 Reinforcement of a Connection with a Tensile Force Component Perpendicular to the Grain

Standard text (Sect. 10.1.4.2):

- *The reinforcement of a connection with a tensile force component perpendicular to the grain (see Fig. 8) may be designed for a tensile force $F_{t,90,Ed}$:*

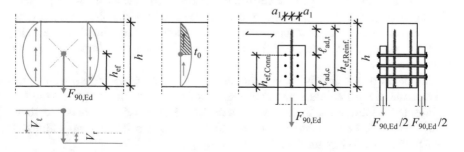

Fig. 8 Reinforced cross-connection: distribution of shear stresses (left), reinforcement (right)

$$F_{t,90,Ed} = \left[1 - 3 \cdot \alpha^2 + 2 \cdot \alpha^3\right] \cdot F_{90,Ed} \qquad (15)$$

where
$\alpha = h_{ef,Conn.}/h$ *(see Fig. 8).*
$h_{ef,\,Conn.}$, *is the effective depth of the connection (see Fig. 8).*
- *Where the effective depth of the reinforcement $h_{ef,Reinf.}$ (see Fig. 8) is smaller than 0,7 · h, measured from the loaded beam edge, Formula (15) should be satisfied at the tip respectively edge of the reinforcement facing the unloaded beam edge.*
- *Where the tensile force $F_{t,90,Ed}$ (according to Formula 15) is carried by internal dowel-type reinforcement, only one row of the internal dowel-type reinforcement at a distance parallel to the grain, $a_{3,c,}$ from the edge of the connection should be considered (see Fig. 8).*
- *The plane reinforcement, e.g. panels or laminations, should be glued to the member according to Fig. 8, with the following limits:*

$$0,25 \leq \frac{b_r}{\ell_{ad}} \leq 0,5 \qquad (16)$$

with

b_r *the width of the plane reinforcement;*
ℓ_{ad} *the relevant effective anchorage length (see Fig. 8).*

Depending on the type of product and type of application (e.g. screw press gluing), a width b_r, exceeding the upper limit given in Formula (16), may be required. The width b_r applied in Formula (4) should not exceed the upper limit given in Formula (16).

Background of the clauses given above:

The tensile force perpendicular to the grain, $F_{t,90,Ed}$, is the resultant of the tensile stresses perpendicular to the grain on the plane defined by the distance of the loaded edge to the centre of the most distant fastener, $h_{ef,Conn.}$ (see [34] for example). According to beam theory, the component of the connection force perpendicular to the grain results in a step in the shear force distribution. The tensile force perpendicular to the grain, $F_{t,90,Ed}$, is determined by integration of the shear stress in the area between the row of fasteners under consideration and the unloaded edge. The term in brackets in Formula (15) is the result of this integration; a derivation can be found in [35].

The depth of the reinforcement should be sufficient enough to avoid moving the location of tensile failure perpendicular to the grain from the connection to the tip/edge of the reinforcement. In analogy to the experiences and rules for connections with a tensile force component perpendicular to the grain ([2, 3]), no additional verification is necessary for the relationship $h_{ef,Reinf.}/h > 0,7$.

The distance between the dowel-type reinforcement and the connection is limited to take into account the limited distribution length of the tensile stresses perpendicular to the grain outside the connection. The same is true for the applicable width of reinforcement panels. The recommended minimum width of the reinforcement is based on the same conditions as described in Sect. 7.4. With respect to placing the reinforcement as close as possible to peak stresses, it is recommended to use the minimum possible spacing $a_{3,c}$ between the dowel-type reinforcement and fastener and to place the dowel-type reinforcement between the fasteners of a connection.

7.7 Reinforcement of Bolted/Dowelled Connections

Standard text (Sect. 10.5.1.1):

- *Where splitting of the timber is prevented through sufficient reinforcement perpendicular to the grain, the effective number of fasteners according to formula (8.34), i.e. determination of n_{ef} in [2], may be taken as $n_{ef} = n$.*
- *The characteristic tensile force in the reinforcement may be taken as $F_{t,90,Ek}= 0{,}3 \cdot F_{v,Rk}$, with $F_{v,Rk}$ determined for one bolt/dowel and one shear plane according to Formulae (8.6)–(8.7) and (8.9)–(8.13) in [2].*
- *The verification of block shear applies (Annex A in [2]) (Fig. 9).*

Background of the clauses given above:

The load-carrying capacity per dowel in connections with multiple dowels placed in rows parallel to the grain and loaded by a load component parallel to the grain is smaller than the load-carrying capacity of a connection with one single dowel. This reduction in the load-carrying capacity in connections with multiple dowel-type fasteners is mainly the result of premature splitting of the timber in the direction of the rows of dowels. The effective number of dowels, according to Formula (8.34) in [2], is based on [36].

Splitting may be prevented by reinforcing the connection area with, for example, self-tapping screws or wood-based panels. In [37], it is demonstrated that in connections with sufficient reinforcement between the dowels, the timber does not split and the effective number n_{ef} equals the actual number n of dowels in one row. With reference to [38], it is stated in [37] that timber splitting is prevented if the

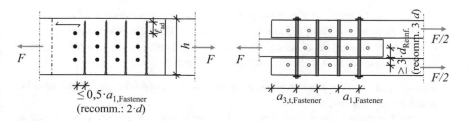

Fig. 9 Reinforced bolted/dowelled connection: arrangement and distance requirements

axial load-carrying capacity, $F_{ax,R}$, of each screw exceeds 30% of the lateral load-carrying capacity, $F_{v,R}$, of one dowel and one shear plane. The axial load-carrying capacity of the dowel-type reinforcement is determined by the effective anchorage length between the outermost row of bolts/dowels and the tip of the dowel-type reinforcement. In practice, a distance, between the dowel/bolt and the dowel-type reinforcement, of twice the diameter of the reinforcement has proven sufficient for a safe insertion of the reinforcement. According to [38], splitting is also prevented if the dowel-type reinforcement is placed at a larger distance, e.g. half the distance between two adjacent fasteners. The dowel-type reinforcement should be placed at a minimum possible distance, $a_{4,c}$, to the corresponding shear plane to reduce the distance to the area of peak splitting stresses on the timber adjacent to the shear plane.

7.8 Reinforcement of Members with Compression Stresses Perpendicular to the Grain

Standard text (Sect. 8.1.5.2):

- *This subclause applies for*

  - *members made from softwoods;*
  - *with reinforcements to carry compression stresses perpendicular to the grain;*
  - *either by fully threaded screws or screwed-in threaded rods.*

The screws or rods should be

  - *applicable for the corresponding timber product and service class of the reinforced timber member;*
  - *evenly distributed over the reinforced contact area;*
  - *applied at an angle between screw or rod axis and grain direction of $45° \leq \alpha \leq 90°$;*
  - *applied at an angle between screw or rod axis and contact surface of $90°$;*
  - *applied with their heads flush with the contact area.*

The contact area should have

  - *adequate stiffness (e.g. a steel plate of adequate thickness) and evenness to prevent penetration of the screw or rod heads into the contact member;*
  - *adequate rotational capacity, where necessary, to provide an equal distribution of the compression force over all screws or rods;*

The contact width at the tip of the reinforcement should be equal to the member width b (see Fig. 10).
For such reinforcement, the characteristic resistance of the reinforced contact area should be taken as the minimum value from Formula (17):

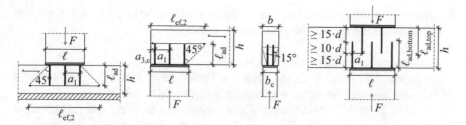

Fig. 10 Reinforced support areas: member on continuous support (left) and discrete support (middle left), cross-section (middle right), transmission of perpendicular-to-grain forces (right)

$$F_{c,90,Rk} = \min \left\{ \begin{array}{c} k_{c,90} \cdot b_c \cdot \ell_{ef,1} \cdot f_{c,90,k} + n \cdot \min\{F_{ax,\alpha,Rk}; F_{b,Rk}\} \\ b \cdot \ell_{ef,2} \cdot f_{c,90,k} \end{array} \right. \qquad (17)$$

where

$k_{c,90}$	*is according to 6.1.5.1 in [2];*
b_c	*is the contact width in mm (see Fig. 10);*
$\ell_{ef,1}$	*is the effective contact length, in mm, parallel to the grain (according to 6.1.5 in [2]); for $\alpha < 90°$, $\ell_{ef} = \ell$, in mm;*
$n = n_0 \cdot n_{90}$	*is the number of fully-threaded screws or rods applied for reinforcement;*
n_0	*is the number of fully-threaded screws or rods arranged in a row parallel to the grain;*
n_{90}	*is the number of fully-threaded screws or rods arranged in a row perpendicular to the grain;*
$F_{ax,\alpha,Rk}$	*is the characteristic withdrawal capacity at the given angle to the grain (according to 8.7.2 in [2] or Technical Assessment);*
$F_{b,Rk}$	*is the characteristic capacity of the screw in axial compression, in N (see below or Technical Assessment);*
b	*is the member width, in mm (see Fig. 10).*
$\ell_{ef,2}$	*is the effective distribution length parallel to the grain in the plane defined by the screw or rod tips (see Fig. 10);*

with

$$\ell_{ef,2} = \ell_{ad} + (n_0 - 1) \cdot a_1 + \min\{\ell_{ad}; a_{3,c}\} \text{ for end supports} \qquad (18)$$

$$\ell_{ef,2} = 2 \cdot \ell_{ad} + (n_0 - 1) \cdot a_1 \text{ for intermediate supports} \qquad (19)$$

where

ℓ_{ad}	*is the penetration length of the threaded part of the screw or rod in the timber member, in mm (see Fig. 10);*
$a_1, a_{3,c}$	*are the spacing parallel to grain and end distance, in mm.*

- *Minimum spacing and end and edge distances should be taken from* Table 8.6 in [2] *or from the European Technical Assessment.*

Background of the clauses given above:

Structural details in which the timber is loaded in compression perpendicular to the grain are very common, e.g. beam supports or sills/sole plates. The combination of high loads to be transferred over localized areas and low capacities in compression perpendicular to the grain can make it difficult to meet the associated verifications. Fully threaded, self-tapping screws or screwed-in threaded rods are a means to improve the stress dispersion in the timber. Partially threaded screws do not work for this application. The main developments in this field have been presented in [39].

In contrast to the design approach applied for reinforcement to carry tensile stresses perpendicular to the grain, the load-carrying capacity of a reinforced support can be determined on the assumption of an additive coaction between the timber under compression stresses perpendicular to the grain and the screws/rods under compression. This assumption is valid if certain deformations of the loaded edge are accepted. This also explains why the effective number of fasteners, n_{ef}, does not apply in this application. In addition, verification of the compression resistance of a fully threaded screw (pushing-in or buckling) is necessary (see subsequently). Typically, buckling of the screw limits the ratio of effective length to diameter (ℓ/d). Finally, it should be verified that the compression capacity perpendicular to the grain of the timber is not exceeded at the screw tips (transition between reinforced and unreinforced section), in a plane defined by an effective length, $\ell_{ef,2}$. The effective length is not to be interpreted as the support length, hence the factor $k_{c,90}$ is not applicable in this verification [39]. At the screw tips, the failure behaviour under compression stresses perpendicular to the grain is characterized by transverse deformation (elongation) over the width of the member. Over the supports, this deformation is prevented by the bearing material. The angle of stress distribution applied to determine the effective length, $\ell_{ef,2}$, used for verification at the screw tips may be taken as 45°, measured from the screw heads. The definition of stress distribution has changed over the years (linear load distribution under 45°, measured from the edge of the steel plate [40]; exponential load distribution, measured from the edge of the steel plate [39]; linear load distribution under 45°, measured from the screw heads, e.g. [41]), hence different approaches can be found in the literature. For longer reinforcing elements, the assumed load distribution angle of 45°, measured from the screw heads, delivers results on the safe side. In case of support conditions with 45° ≤ α < 90°, the load distribution angle will decrease with a decrease in the angle between load and grain. This is compensated for by the use of $f_{c,90,k}$ (and not $f_{c,\alpha,k}$) for all angles between load and grain, which delivers results on the safe side. The load distribution perpendicular to the grain over the width of the member is smaller. If necessary, e.g. in block-glued members, the angle of load distribution perpendicular to the grain can be taken as 15° [42]. To guarantee a homogeneous transfer of shear into the reinforced support area, the contact width at the tip of the reinforcement, determined by the angle of load distribution perpendicular to the grain, should equal the member width (see Fig. 10).

The compression force must be evenly distributed across all the screws and the compression stresses at the screw heads have to be absorbed by the bearing material. These two requirements can only be met if a hard bearing material is used. This can be realized in the form of a hard intermediate layer from, say, steel, with adequate thickness and thus capable of transferring the load uniformly. The screws shall be equally distributed all over the bearing area and the screw heads shall be on one line with the surface of the timber member.

The distance requirements are the same as for screws in tension. It is not necessary to take into account the effective number of screws, n_{ef}.

- *The contact material (e.g. steel plate) should be designed for the load introduced by the screw head. The thickness of the steel plate, t, may be assumed adequate if Formula (20) is satisfied:*

$$t \geq \max\left(5, 0; \quad 1, 45 \cdot \sqrt{\frac{F_{c,\alpha,Ed}}{f_{y,d}}}\right) \tag{20}$$

where

t	*is the thickness of the steel plate, in mm;*
$F_{c,\alpha,Ed}$	*is the design compression force in one screw or rod, in N;*
$f_{y,d}$	*is the design yield strength of the steel plate, in N/mm².*

- *Where rotation of the member results in indentation of the member due to the stiff contact material (e.g. steel plate), it is recommended to increase the rotational capacity (by means of, for example, an elastomeric bearing material) if the following limit is exceeded:*

$$\Delta w = \varphi \cdot \ell/2 \geq 1, 0 \, \text{mm} \tag{21}$$

where

Δw	*is the relative deformation of the member from rotation at the edge of the contact;*
φ	*is the rotation of the member at the support at u_{fin} (see 2.2.3(3) in [2]);*
ℓ	*is the contact length, in mm (see Fig. 10).*

- *Screws or rods driven into the top and bottom of a member may overlap. The characteristic resistance may be determined, according to the first part of Formula (17), if*

  - *the contact areas and the screws or rods are axially symmetrical on both opposite sides; and*
  - *the screws or rods overlap at least 10 · d, where d is the screw or rod diameter; and*
  - *the distance between the screw or rod tips and the opposite contact area is at least 15 · d (see Fig. 10).*

- *Reinforcement with glued-in rods may be designed in analogy to the above given clauses.*
- *Section 7.2 (Effects of moisture content changes) applies.*
- *At supports where the support force is not fully transferred to the contact area (e.g. reinforced support area) and at supports where the support force is transferred via connections fastened to the end grain, reduction in the total shear force (6.1.7 in [2]) should not be applied.*

Note to the reader: In the draft standard, the last bullet point is placed in Sect. 6.1.7 in [2]. For reasons of representation, the introduction of this information in a separate section has been ignored in this contribution.

A hard intermediate layer (e.g. steel plate) should, amongst others, be designed for the bending stresses due to the forces induced by the screw heads. This verification can be converted into an equation which sets into a relation the force in the dowel-type reinforcement in compression and the yield strength of the steel plate.

The requirement of adequate rotational capacity can be met by placing a softer interlayer (e.g. elastomeric bearing) between the steel plate and the member below. Not all types of support require high rotational capacity, e.g. the mid-supports of a 2-span beam. The given threshold value is based on practical experience.

Reinforcement driven in from both edges of the beam to enable the transfer of compression perpendicular to grain stresses through the timber member has been introduced by [43], and subsequently integrated into the Technical Approvals [44]. The validity of the approach, the minimum overlap, and the minimum distance between the screw tips and the opposite contact area have been numerically and experimentally verified in [45]. With respect to facilitated stress transfer between the screws applied from the opposite edges, it is recommended to use the minimum possible spacing and an alternating arrangement of the screws over the contact area.

Information on the verification of fully threaded screws or screwed-in threaded rods in axial compression is given below. In [39] a calculation model to determine the stiffness of reinforced beam supports is proposed.

- *The characteristic compression resistance (pushing-in or buckling), $F_{c,\alpha,\mathrm{Rk}}$, may be simplified as:*

$$F_{c,\alpha,Rk} = \min\{F_{ax,\alpha,Rk}; F_{b,Rk}\} \tag{22}$$

where

$F_{ax,\alpha,\mathrm{Rk}}$ *is the characteristic withdrawal capacity according to Formula 8.39 respectively 8.40 in [2];*

$F_{b,Rk} = 1,18 \cdot k_c \cdot N_{pl,k}$ *is the characteristic load-carrying capacity of the screw in axial compression;*

with
values k_c given in Table 2
and

Table 2 Reduction factors k_c for buckling of screws ($\rho_k \geq 350$ kg/m^3)

	Angle α between screw axis and grain	
Characteristic value of yield strength of steel	$\alpha = 90°$	$\alpha = 0°$
$f_{y,k} = 1000$ N/mm^2	$k_c = 0{,}6$	$k_c = 0{,}5$
$f_{y,k} = 800$ N/mm^2 *(e.g. hot dip galvanized steel)*	$k_c = 0{,}65$	$k_c = 0{,}55$
$f_{y,k} = 500$ N/mm^2 *(e.g. stainless steel)*	$k_c = 0{,}75$	$k_c = 0{,}65$

NOTE for characteristic values of yield strength of steel in between any two of the specified values, k_c can be determined by linear interpolation. For angles α between the specified values, k_c can be determined by linear interpolation

$$N_{pl,k} = \pi \cdot \frac{d_1^2}{4} \cdot f_{y,k}$$

where

d_1 *is the inner thread diameter;*
$f_{y,k}$ *is the characteristic yield strength of the screw.*

The characteristic compression resistance of a fully threaded screw is taken as the smaller of the pushing-in and buckling capacities. The pushing-in capacity is considered equal to the withdrawal capacity of the fully threaded screw. The full equations to determine the buckling capacity of fully threaded screws have been determined in [39]. The simplification of the equations proposed above has been developed in [46]. Bejtka and Blaß [39] also proposes buckling capacities for the application of fully threaded screws with clamped heads in a steel plate. This application necessitates the countersinking of the steel plate in the form of screw heads, such that the surfaces of the screw heads are flush with the lower steel plate surface. In practice, this can be a challenge due to the necessity of exact manufacturing together with the multitude of forms of screw heads available in the market.

8 Conclusion and Outlook

Despite the current lack of design approaches in Eurocode 5 and comparable standards on structural timber design, reinforcement for stresses perpendicular to the grain with fully-threaded self-tapping screws, glued-in or screwed-in threaded rods, or glued-on plywood and LVL can be considered state-of-the art in timber engineering practice. For numerous applications, design procedures are available which have already been clarified to an extent to satisfy safety requirements and engineering needs. These are currently being prepared for introduction in the next generation of Eurocode 5. This contribution gives an overview of these developments.

Future research and development should focus on better quantification of the reinforcing effect of inclined dowel-type reinforcement, clearer determination of stiffness properties of the reinforcement in the timber before cracking, as well as a better understanding and quantification of the potentially harmful effect of reinforcement restricting the free shrinkage or swelling of the timber.

Acknowledgements The technical input and comments from the following members of CEN/TC 250/SC 5/WG 7 and PT SC5.T1 are thankfully acknowledged: Alfons Brunauer† (AT), Prof. Robert Jockwer (SE), Harald Liven (NO), Prof. João Negrão (PT), Prof. Erik Serrano (SE), and Dr. Tobias Wiegand (DE).

References

1. Harte A, Dietsch P (eds) (2015) Reinforcement of timber structures. A state-of-the-art report. Shaker Verlag, Aachen. ISBN 978-3-8440-3751-7
2. EN 1995-1-1:2004, Eurocode 5: design of timber structures—part 1-1: general—common rules and rules for buildings, +AC (2006) +A1 (2008) +A2 (2014). CEN European committee for standardization, Brussels
3. DIN EN 1995-1-1/NA:2013. National annex—nationally determined parameters—Eurocode 5: Design of timber structures—part 1-1: general—common rules and rules for buildings. DIN Deutsches Institut für Normung e. V., Beuth Verlag GmbH, Berlin
4. ÖNORM B 1995-1-1:2015-06. Eurocode 5: design of timber structures—part 1-1: general—common rules and rules for buildings—national specifications for the implementation of ÖNORM EN 1995-1-1, national comments and national supplements. ASI Austrian Standards Institute, Vienna
5. Dietsch P, Winter S (2012) Eurocode 5—future developments towards a more comprehensive code on timber structures. Struct Eng Int 22(2):223–231
6. CEN/TC 250 N1239 (2014) Position paper on enhancing ease of use of the structural Eurocodes. CEN/TC 250 Document Ref. N1239, Brussels
7. Kleinhenz M, Winter S, Dietsch P (2016) Eurocode 5—a halftime summary of the revision process. In: Proceedings of the world conference on timber engineering WCTE 2016, Vienna, Austria
8. Harte A, Jockwer R, Stepinac M, Descamps T, Rajcic V, Dietsch P (2015) Reinforcement of timber structures—the route to standardization. In: Proceedings 3rd international conference on structural health assessment of timber structures SHATIS, Wroclaw, ISBN 978-83-7125-255-6
9. https://www.cen.eu/work/products/TS/Pages/default.aspx
10. Dietsch P, Brandner R (2015) Self-tapping screws and threaded rods as reinforcement for structural timber elements—a state-of-the-art report. Constr Build Mater 97:78–89
11. Steiger R, Serrano E, Stepinac M, Rajcik V, O'Neill C, McPolin D, Widmann R (2015) Strengthening of timber structures with glued-in rods. Constr Build Mater 97:90–105
12. Brüninghoff H, Schmidt C, Wiegand T (1993) Praxisnahe Empfehlungen zur Reduzierung von Querzugrissen bei geleimten Satteldachbindern aus Brettschichtholz. Bauen mit Holz 95(11):928–937
13. Ranta-Maunus A, Gowda SS (1994) Curved and cambered glulam beams—part 2: long term load tests under cyclically varying humidity. VTT Publications 171, Technical Research Centre, Espoo
14. Gustafsson PJ (1995) Notched beams and holes in glulam beams. In: Blaß HJ et al (eds) Timber Eng STEP 1. Centrum Hout, Almere
15. Blaß HJ, Ehlbeck J, Kreuzinger H, Steck G (2004) Erläuterungen zu DIN 1052:2004-08. Bruderverlag, Karlsruhe
16. EN 14080:2013. Timber structures—glued laminated timber and glued solid timber—requirements. CEN European committee for standardization, Brussels

17. EN 1993-1-8:2005 + AC:2009. Eurocode 3: design of steel structures—part 1-8: design of joints. CEN European committee for standardization, Brussels
18. Blaß HJ, Steck G (1999) Querzugverstärkungen von Holzbauteilen: Teil 1–Teil 3. Bauen mit Holz, vol 101, Issue 3, pp 42–46, Issue 4, pp 44–49, Issue 5, pp 46–50
19. Frühwald E, Serrano E, Toratti T, Emilsson A, Thelandersson S (2007) Design of safe timber structures—how can we learn from structural failures in concrete, steel and timber? Report TVBK-3053, Div. of Struct. Eng, Lund University
20. Frese M, Blaß HJ (2011) Statistics of damages to timber structures in Germany. Eng Struct 33(11):2969–2977
21. Dietsch P, Winter S (2018) Structural failure in large-span timber structures: a comprehensive analysis of 230 cases. Struct Saf 71:41–46
22. Wallner B (2012) Versuchstechnische Evaluierung feuchteinduzierter Kräfte in Brettschichtholz verursacht durch das Einbringen von Schraubstangen, Master thesis, Institute of Timber Engineering and Wood Technology, Graz University of Technology
23. Dietsch P (2017) Effect of reinforcement on shrinkage stresses in timber members. Constr Build Mater 150:903–915
24. Blaß HJ, Krüger O (2010) Schubverstärkung von Holz mit Holzschrauben und Gewindestangen. Karlsruher Berichte zum Ingenieurholzbau, Band 15, Universitätsverlag Karlsruhe
25. BVPI (2010) Spannungsnachweise bei Satteldachträgern aus Brettschichtholz. Technische Mitteilung 06/011 der Bundesvereinigung der Prüfingenieure für Bautechnik e.V., Berlin
26. Dietsch P, Winter S (2015) Untersuchung von nicht im Eurocode 5 geregelten Formen von Satteldachträgern im Hinblick auf den Nachweis der Querzugspannungen. Forschungsbericht, Lehrstuhl für Holzbau und Baukonstruktion, Technische Universität München
27. Henrici D (1990) Beitrag zur Bemessung ausgeklinkter Brettschichtholzträger. Bauen mit Holz 92(11):806–811
28. Jockwer R (2014) Structural behaviour of glued laminated timber beams with unreinforced and reinforced notches. Dissertation, IBK Bericht Nr. 365, ETH Zurich
29. Aicher S, Höfflin L (2009) Glulam beams with holes reinforced by steel bars. CIB-W18/ 42-12-1. In: Proceedings of the international council for research and innovation in building and construction, Working commission W18—timber structures, Meeting 42, Zürich
30. Aicher S (2011) Glulam beams with internally and externally reinforced holes—test, detailing and design. CIB-W18/ 44-12-4. In: Proceedings of the international council for research and innovation in building and construction, Meeting 44, Alghero
31. Danzer M, Dietsch P, Winter S (2017) Round holes in glulam beams arranged eccentrically or in groups. INTER/ 50-12-6, International Network on Timber Engineering Research INTER, Meeting 4, Kyoto
32. Kolb H, Epple A (1985) Verstärkungen von durchbrochenen Brettschichtholzbindern. Schlussbericht zum Forschungsvorhaben I.4 – 34810, Forschungs- und Materialprüfungsanstalt Baden-Württemberg, Stuttgart
33. Blaß HJ, Bejtka I (2004) Reinforcements perpendicular to the grain using self-tapping screws. In: Proceedings, 8th world conference on timber engineering WCTE 2004, Lahti
34. Ehlbeck J, Görlacher R, Werner H (1991) Empfehlung zum einheitlichen genaueren Querzugnachweis für Anschlüsse mit mechanischen Verbindungsmitteln. Bauen mit Holz 93(11):825–828
35. Dietsch P, Kreuzinger H, Winter S (2019) Holzbau. In: Zilch K et al (eds) Handbuch für Bauingenieure. Springer, Berlin
36. Jorissen AJM (1998) Double shear timber connections with dowel type fasteners. Dissertation, Delft University of Technology
37. Bejtka I, Blaß HJ (2005) Self-tapping screws as reinforcements in connections with dowel-type fasteners. CIB-W18/38-7-4. In: Proceedings of the international council for research and innovation in building and construction, Working commission W18—timber structures, Meeting 38, Karlsruhe

38. Schmid M (2002) Anwendung der Bruchmechanik auf Verbindungen mit Holz. Dissertation, Universität Karlsruhe (TH)
39. Bejtka I, Blaß HJ (2006) Self-tapping screws as reinforcements in beam supports. CIB-W18/39-7-2. In: Proceedings of the international council for research and innovation in building and construction, Working commission W18—timber structures, Meeting 39, Florence
40. Bejtka I (2003) Querzug- und Querdruckverstärkungen Aktuelle Forschungsergebnisse. In: Proceedings Ingenieurholzbau - Karlsruher Tage, Bruderverlag, Karlsruhe
41. ETA-12/0114 (2017) Self-tapping screws for use in timber structures. SPAX International GmbH & Co. KG, ETA Denmark
42. EN 1995-2:2004. Eurocode 5: Eurocode 5: design of timber structures—part 2: bridges. CEN European committee for standardization, Brussels
43. Watson C, van Beerschoten W, Smith T, Pampanin S, Buchanan AH (2013) Stiffness of screw reinforced LVL in compression perpendicular to the grain. CIB-W18/46-12-4. In: Proceedings of the international council for research and innovation in building and construction, Working commission W18—timber structures, Meeting 47, Vancouver
44. DIBt Z-9.1-519 (2014) Allgemeine bauaufsichtliche Zulassung, SPAX-S Schrauben mit Vollgewinde als Holzverbindungsmittel. Deutsches Institut für Bautechnik, Berlin
45. Dietsch P, Rodemeier S, Blaß HJ (2019) Transmission of perpendicular to grain forces using self-tapping screws. INTER/52-7-10, International network on timber engineering research INTER, Meeting 6, Tacoma
46. Jockwer R (2016) Simplification of the design approach for buckling failure of reinforcement in compression. Short Report, ETH Zurich

Seismic Reinforcement of Traditional Timber Structures

Chrysl A. Aranha and Jorge M. Branco

Abstract Every year earthquakes cause damage and destroy a sizeable portion of the building stock across the globe. Among traditional constructions, those built with timber are considered the most effective earthquake-resistant structures, provided that the continuity in the load path is not compromised, the joints are intact, and moisture-induced problems are kept at bay. However, the high costs and difficulties involved in the execution of interventions that meet the safety requirements prescribed by current building codes act as a deterrent to the continued use and reuse of these structures. Therefore, it is important to identify the inherent seismic-resistant features, as well as the deficiencies, of traditional timber constructions and to review the various strengthening, retrofitting and upgrading measures that have been developed to enhance the safety of such structures. The effectiveness of different strengthening techniques has been proven on the basis of results from experimental tests carried out on components of timber structures, such as joints and beams, full-scale shear walls, roof trusses and floor slabs. The successes and failures of past interventions also play an instrumental role in identifying effective and economical strengthening solutions for traditional timber structures.

1 Introduction

The versatility of wood as a building material led to its widespread use in construction for thousands of years. Prior to the advent of concrete and steel, the choice of timber as a beam element was only natural because of its ability to withstand both tensile and compressive loads. However, due to durability issues, its vulnerability to fire [1], and the commonly held perception of the impermanence of timber, monuments and structures of importance built wholly of timber are relatively uncommon. The

C. A. Aranha
Civil Engineering Department, ISISE, University of Minho, Braga, Portugal

J. M. Branco (✉)
Department of Civil Engineering, ISISE, University of Minho, Guimarães, Portugal
e-mail: jbranco@civil.uminho.pt

© RILEM 2021
J. Branco et al. (eds.), *Reinforcement of Timber Elements in Existing Structures*,
RILEM State-of-the-Art Reports 33,
https://doi.org/10.1007/978-3-030-67794-7_7

133

tradition of building houses entirely of timber is found in richly forested regions characterised by cold climate. In sub-tropical and tropical countries, the use of timber members is generally restricted to floors and roofs.

1.1 Seismic Response of Traditional Timber Structures

Log house construction is one of the foremost practices of building with wood, wherein logs are simply stacked and overlapped at their notched corners to create box-like structures [2]. The lateral load resistance of these structures depends on the efficiency of the carpentry joints at the cross-wall corners and, to some extent, on the friction between log surfaces in contact [3]. However, shrinkage of logs and creep might result in the loosening of these joints over time, which renders them vulnerable to shear failure during seismic events. Among countries with high seismic risks, traditional log structures are seen in Turkey [4].

China has a long-standing tradition of building timber frame structures connected, using mortise and tenon joints, with various intricate bracketing systems. The Yingxian pagoda, which dates back to the eleventh century C.E, has managed to withstand five major earthquakes till date without undergoing any major damage [5]. Studies have shown that friction damping and energy dissipation resulting from the workings of the mortise and tenon joints and the brackets play an integral role in the seismic performance of such structures [6]. In Japan, timber has been used in the construction of Shinto shrines, Buddhist temples and houses. Traditional timber structures are of the post-and-beam type with mortise and tenon connections secured using wooden wedges and pegs. Being an earthquake-prone country, a typical timber system is designed with inherent damping mechanisms and can act as a flexible structure with a certain degree of flexure and capability to sway under external lateral loads [7].

Mixed timber-masonry constructions are found in many parts of the world, where they have developed as vernacular systems. The Alpide belt, which is the second most seismically active region of the world and is responsible for about 15% of the world's earthquakes, extends from Azores in Portugal through Spain, Italy, Greece, Turkey and northern India, and then heads south into Sumatra, Indonesia [8]. It is interesting to note that variations of earthquake-resistant timber-laced masonry houses have been developed as vernacular constructions in parts of countries, such as Italy, Greece, Turkey, Pakistan, India, and Nepal, along the seismic belt. This trend is especially clear in Fig. 1, which shows the different vernacular timber-laced masonry systems superposed on the map of the seismic zones in India marked according to IS I893 [9]. These zones are categorized based on the levels of intensities sustained during past earthquakes, with areas located in zone V having the highest expected level of quakes.

In highly seismic areas of South America, vernacular timber-laced masonry constructions are termed bahareque (Colombia, Venezuela, San Salvador and El Salvador), taquezal (Nicaragua) or quincha (Peru). It is noteworthy that timber-laced

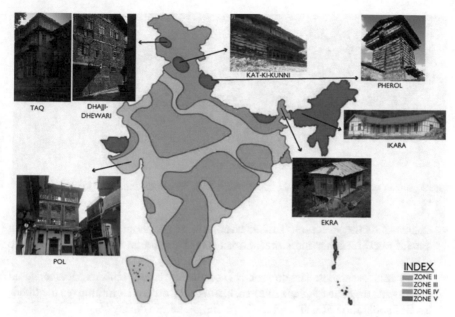

Fig. 1 Map of seismic zones in India showing the distribution of vernacular timber-laced masonry systems

masonry systems were also specifically developed as aseismic constructions in post-earthquake reconstruction in Portugal (Pombalino) and Italy (Casa baraccata) in the eighteenth Century CE [10].

1.2 Earthquake-Resistant Features of Traditional Timber Structures

The seismic performance of traditional timber structures is positively affected by regularity in plan and elevation, redundancy, minimizing the number and extent of openings, and keeping the storey height low. The seismic resistance of traditional timber structures can be attributed to their high strength-to-weight ratio, redundancy and the so-called pseudo-ductility exhibited by their joints [11, 12]. The main lateral load-resisting elements of timber structures are horizontal floor and roof diaphragms and vertical shear walls.

Traditional timber-laced masonry constructions have specific features that are responsible for the improved seismic performance of these non-engineered structures. The timber elements present in these structures contribute to their seismic resistance in the following ways:

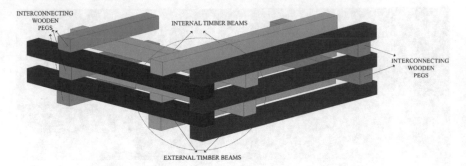

Fig. 2 Timber elements of the Kat-ki-kunni system with the corner-reinforced joint marked

- Lightness of the structure: Timber-laced masonry structures, being lighter than unreinforced masonry structures, attract lower horizontal forces during seismic events.
- Corner reinforcement: The presence of corner-reinforced joints restricts torsional movements and out-of-plane failures. Corner reinforcement minimizes the likelihood of separation of walls, which is a common failure during seismic events. The Kat-ki-kunni houses, characterized by their reinforced timber corners, sustained only minor damage during the Kangra earthquake of 1905 [13]. In Fig. 2, a portion of the timber lattice of this system can be seen.
- Box behaviour: Timber ring beams at the plinths, lintels, floor and roof levels tie the walls together and help the structure to behave as a single unit [14]. During an earthquake, the walls loaded in their weak direction can transfer loads to the adjacent walls loaded in their strong direction through the ring beam.
- Shear crack prevention: Horizontal and vertical timber members confine the masonry and improve its bearing capacity. Diagonal timber members, which function as cross-bracing, prevent the spreading of X-shaped shear cracks that are typical of in-plane shear failure in masonry walls [15] (see Fig. 3).
- Improved energy dissipation: The horizontal and vertical timber members ensure energy dissipation along two slip planes [16].

2   Considerations for Seismic Strengthening of Traditional Timber Structures

Even though traditional timber structures have been known to perform well during past earthquakes, significant damage can be incurred by these structures owing to various factors. One of these factors is the presence of deteriorated timber members due to lack of maintenance, leading to moisture problems and insect infestation; others include discontinuities in the timber frame structure, insecure timber members, insufficient panel bracing, heavy roofing material and failure of connections due to the inadequacy of joints [17–20].

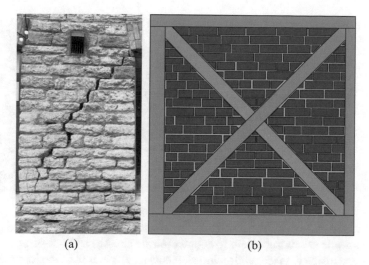

(a) (b)

Fig. 3 **a** Diagonal shear crack in unreinforced masonry wall and **b** infill-frame wall with diagonal timber braces

2.1 Considerations for Timber Components

Failure of timber elements under tension, bending and shear is brittle and should be avoided. The behaviour of traditional timber structures is, for the most part, governed by their joints [21–23]. Hence, the occurrence of brittle failure in timber structures can be prevented by designing the structures such that their members remain elastic and their joints undergo ductile failure. In fact, post-earthquake assessment studies of damaged timber buildings indicate that one of the primary causes of their failure is the inadequacy of their joints [24]. Under seismic loads, the in-plane stiffness of floors and roofs and the ability of the roofs to transfer loads to the walls below play a decisive role in the seismic performance of timber structures. Hence, the considerations for seismic strengthening of these individual components are also addressed in this chapter.

2.1.1 Considerations for Timber Joints

Various types of carpentry joints evolved from historic timber structures because of traditional practices or carpenters' intuition based on empirical rules [25]. Traditional carpentry joints rely heavily on contact and transfer loads by compression and friction [26] and usually have some moment capacity [27]. Although historic timber joints may not comply with the safety levels imposed by current building codes, interventions are only carried out when there is an increase in the magnitude of the loads or when the members show signs of structural distress.

Fig. 4 A broken wooden
dowel that no longer secures
the joint between a collar
beam and a rafter

During seismic events, the reversal of loads necessitates the use of metal parts or wooden dowels as reinforcement to counter tensile stresses and out-of-plane movement. Any damage to these reinforcing elements results in loss of the joint functionality. In Fig. 4, the fracture of the wooden dowel, whose main purpose is to prevent the out-of-plane movement of the collar beam and rafter, results in loss of functionality of the joint.

2.1.2 Considerations for Timber Floors and Roofs

The main aim of reinforcing timber floors in traditional structures is to increase the in-plane as well as out-of-plane stiffness and improve the bearing capacity [28, 29]. Under seismic loading, vulnerabilities of timber roofs arise from insufficient in-plane stiffness, unrestrained thrusts, ineffective support conditions, and joints with inadequate capacity [30]. Therefore, seismic strengthening of roof structures has to take into account all the aforementioned considerations.

2.2 Aspects to Be Considered for Seismic Reinforcement Design

The effective design of seismic reinforcement of timber structures depends on the determination of the following parameters before implementation:

- The state of damage on the structure;
- Missing timber elements and/or conceptual design errors;
- Mechanical properties, creep effects, moisture content and residual cross-sections of elements;
- Joint eccentricity, capacity and stiffness;
- Condition of the supports;
- Estimation of stress levels in the structure;

- Load path and distribution;
- Prior interventions and/or additions to the structure, if any, and their influence on the structural performance;
- Design of appropriate shoring to carry out the intervention;
- Impact of the intervention at the local level in terms of compatibility, reversibility, integrity and durability;
- Effect of the intervention on the safety and global stability of the structure;
- Prevention of brittle failure;
- Target displacement levels, energy dissipation, and ductility of the strengthened structure;
- Post-intervention fire performance.

3 Seismic Reinforcement Methods

Reinforcement of timber structures under seismic loads is carried out with the aim of enhancing strength, stiffness, or ductility, and the energy dissipation capacity of specific components and sub-assemblies that will result in an improvement in the seismic performance of the entire structure. It might not be feasible to design an intervention to address all four characteristics since they are interrelated and a deficiency in any one can be offset by an increase in one or more of the others [31].

3.1 Materials Used

The stiffness and strength properties of the material used in the strengthening solution has to be determined in advance and care should be taken not to induce stresses in the original members and failure at the interface. Moreover, the connection between the existing and new elements has to be studied to prevent the concentration of stresses and, possibly, to avoid significant changes in the stiffness of the structure [28].

The materials typically used for the seismic strengthening of timber structures are solid timber, steel, fibre reinforced polymers (FRPs), resins, and engineered wood products (EWPs) like plywood, glulam, laminated veneer lumber (LVL) and cross-laminated timber (CLT) [32]. The preferred material for the strengthening of timber structures has always been timber itself. To ensure maximum compatibility, it is recommended to use timber of the same species and moisture content, as well as of similar values of density, strength and stiffness, to those of the original material. With the development of EWPs, which overcome most of the drawbacks associated with solid timber, their use in seismic performance improvement solutions is gaining ground.

Although steel has always been used in conjunction with timber in traditional structures as straps, ties, clamps, etc., considerations have to be made regarding the risk of exposure to fire and the incursion of moisture [33]. Condensation of moisture at the steel-timber interface can result in corrosion of the steel element and subsequent degradation of the timber. To limit corrosion problems, stainless steel should be used.

The use of concrete as a strengthening solution in timber structures is generally restricted to floors. Considerations for interventions using concrete include keeping the surface of the timber dry and limiting the mass of the composite structure. The great merit of this solution is the large increase in the in-plane stiffness provided by the concrete layer [29].

Depending on the purpose of the intervention, type of loading and environmental conditions, FRPs can also be used to strengthen components of timber structures. Some of their applications include flexural strengthening of beams, improving the in-plane stiffness of floors and strengthening of joints. However, since high temperature has an adverse impact on the mechanical properties of polymer matrices and adhesives, caution must be exercised in their use [34].

Among the listed materials for the strengthening of timber structures, epoxy resins are the most recent addition. Resin-bonded repairs are used in the restoration of beam-ends and column-ends, fissure repairs and strengthening of beams. Resins can be used in conjunction with FRPs or steel rods, where they function as adhesives or can be used as grout to fill damaged sections of timber [35]. In spite of the fact that interventions with resins are minimally invasive and easy to implement on site, there are reservations about using them in structures of importance as these materials have been around for only about fifty years while the timber structures that they are used to reinforce can be more than a century old.

Apart from carefully choosing the strengthening material for the intervention, steps have to be taken towards ensuring the maximum efficiency and durability of the strengthened timber structure. In this regard, the requirements prescribed by the National Research Council in Italy [34] are pertinent. They include:

- Ensuring the stress states in the materials involved are within prescribed limits;
- Careful design and construction of the joints due to the inherent weakness at the interface;
- Protecting the strengthening solution from physicochemical, mechanical and thermal conditions that can weaken its performance by adopting appropriate construction details;
- Executing the intervention with good workmanship and under controlled environmental conditions.

3.2 Reinforcement at the Joint Level

Interventions for strengthening carpentry joints are still strongly based on empirical knowledge; and a heuristic approach is commonly adopted for joint strengthening. Under seismic loads, the contact pressure that is responsible for transmission of loads in timber joints varies and this might result in the disassembly of unreinforced joints. The aim of seismic strengthening of joints is mainly to: (i) improve the post-elastic behaviour of timber joints, (ii) prevent the occurrence of brittle failure, (iii) prevent disassembly of the joint [25]. As glued joints and welded connectors are potential zones for brittle failure, their use in seismic reinforcement is generally avoided [28]. Since traditional timber joints have significant rotational capacity, the use of strengthening solutions that restrict rotational movement is not advised [32].

To ensure a better fit between timber members of a joint and to counter shrinkage effects, wooden wedges can be used [12]. This is not a seismic reinforcement technique per se, but it is useful for the elimination of the gap between the members in contact and can restore the connection's original behaviour [26]. The use of wedges to achieve tight fits in traditional Japanese structures is essential for the proper working of the joints, especially under load reversal [36].

Dowel-type fasteners are commonly used as a quick and efficient way of strengthening timber joints. Traditionally, wooden fasteners were used to carry shear forces in interlocked joints. However, as wooden fasteners also undergo dimensional changes due to shrinkage and are prone to brittle failure, they have been replaced in recent times by metal fasteners [37]. The most commonly used metal fasteners in modern timber structures are self-tapping screws, which eliminate the need for pre-drilling and transfer loads more uniformly to the surrounding wood due to the continuous threads over their lengths. However, their limited potential to plasticize can limit their use as reinforcement in traditional timber structures [38]. The application of self-tapping screws in the strengthening of timber structures has been described in detail in the Chapter "Self-tapping Screws as Reinforcement for Structural Timber Elements" in this RILEM state-of-the-art report.

Among the traditional carpentry joints, the seismic strengthening of notched joints (step joints) has been most extensively studied, since they are commonly found in Mediterranean countries with high seismic risk [39]. The techniques that are commonly used for the reinforcement of notched joints between tie beams and rafters include bolts, clamps and stirrups [26, 40], which are illustrated in Fig. 5.

Bolts are easy to install and they restrain the relative movement of the tie beam and the rafter. Notched joints should be reinforced with bolts whenever feasible because

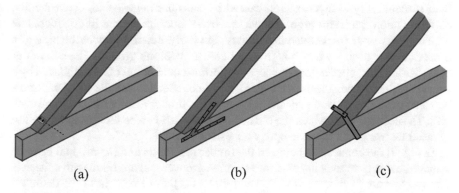

(a)   (b)   (c)

Fig. 5 Reinforcement of rafter-tie beam joint using **a** two transverse bolts, **b** metal stirrups and **c** external clamp (solutions tested in [40])

bolts do not cause an excessive increase in stiffness nor do they modify the behaviour of the joints much. Metal stirrups provide increased in-plane stiffness by securing the connected members and are effective in restricting out-of-plane movement. Clamps restrain the relative movement of the tie beam and the rafter and can be tightened depending on the confining capacity required [26, 40].

3.3 Reinforcement at the Member Level

Deterioration and damage of timber members predispose them to various types of failure under vertical and lateral loads. In the case of decayed timber members, replacement in whole or part depending on the extent of decay is recommended. Prior to replacement, it is essential that the causes of deterioration are identified and necessary measures to eliminate them are taken. The seismic performance of existing timber members can be enhanced by either the addition of side plates or the insertion of flitches. Confinement of timber members using steel is carried out to improve the load-carrying capacity. In the case of traditional timber structures of significance, destruction to the original members of the structure has to be avoided and reversibility is emphasized; post-tensioning techniques are adopted [33].

3.3.1 Replacement

The replacement of entire timber elements is a common practice in Japan, where rebuilding and continuous replacement intervention techniques are typically adopted in the conservation of traditional timber structures [41]. In most other countries around the world, replacement of just the distressed portion of the timber element is preferred to replacement of the entire element.

Typically, in timber structures, substitution of beam ends is common due to their susceptibility to deterioration caused by moisture ingress and, consequently, biodegradation. Moisture problems occur when timber beams are in direct contact with masonry or are embedded in masonry. Similarly, deterioration at the base of a timber column occurs when it is in direct contact with the ground. The column or beam has to be propped before the decayed portion could be cut off and replaced by a suitably fashioned prosthesis (see Fig. 6). It is recommended that the prosthesis must be made of the same wood species as that of the timber member being reinforced [42]. The connection between the old and the new parts can be facilitated by the use of metal fasteners or FRPs and epoxy adhesives [43–45].

In Fig. 7, the connection between the timber prosthesis and the original beam is achieved using steel rods inserted into external grooves that are filled with adhesive. The grooves are then covered with wood fillets [44]. Replacement of the damaged portion of a timber element and ensuring the integrity of the retrofitted element is essential to restoring its load-bearing capacity [43].

Fig. 6 Column end repair with timber prosthesis and metal fasteners at Kipling house, Mumbai

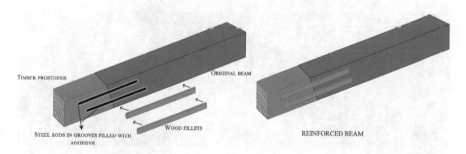

Fig. 7 Beam end repair using timber prosthesis with mechanical reinforcement inserted into external grooves

3.3.2 Addition of Side Plates

A damaged beam can be strengthened by the addition of timber or steel plates on either side of the member [46]. The composite member is then held together using mechanical fasteners. Floor beams can be strengthened in bending by the addition of steel plates on the compression and tension faces, as seen in Fig. 8. The addition of lateral timber or steel plates (see Fig. 9) in the cracked portion of a beam strengthens the otherwise weak section.

The load-bearing capacity of timber columns can be increased by the addition of steel plates on the exterior portion of the columns [47, 48] parallel to the buckling axis,

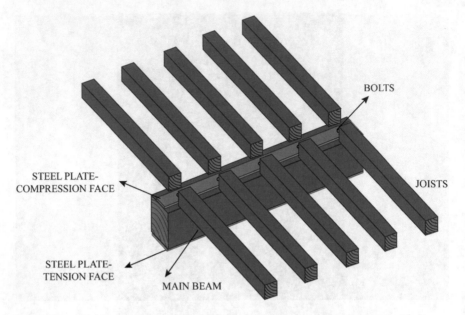

Fig. 8 Strengthening under bending stresses by the addition of steel plates and vertical bolts

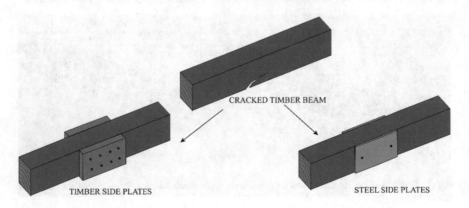

Fig. 9 Lateral reinforcement of a timber beam using side plates

as shown in Fig. 10. The reinforcement strengthens the columns under buckling loads and reduces their vulnerability to splitting failure along the grain caused by bending moments. Depending on the extent of the damage, the member can be reinforced along its length or only in the region of the crack.

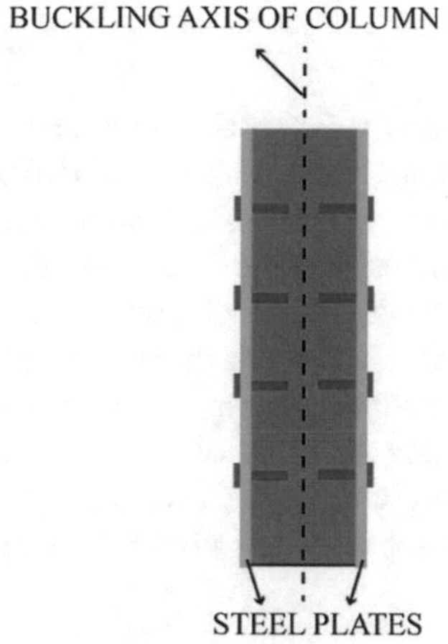

BUCKLING AXIS OF COLUMN

STEEL PLATES

Fig. 10 Lateral confinement of a timber column with steel plates and fasteners (solution tested in [49])

3.3.3   Use of Flitch Plates

This technique consists in removing the decayed end of a timber member, followed by the cutting of a slot along the centre of the member and inserting a flitch into the slot. The member is restored to its original size using suitably dimensioned timber elements placed on either side of the flitch plate and by bolting the composite beam together. The bottom plate of the steel flitch can be concealed using timber inserts [50], as shown in Fig. 11. This reduces the risk of fire associated with exposed steel in timber structures.

The relative stiffness (EI) values determine the load sharing between the steel and timber elements. Flitch plates are generally used in the strengthening of principal rafters and main floor beams [51]. The difficulties involved in aligning the drilled holes through the steel and timber elements and the subsequent bolting together of the member, as well as the requirement of full access to the member, are some of the limitations to the use of this reinforcement technique [52].

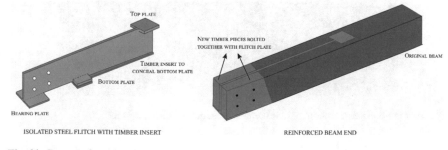

Fig. 11 Beam end repair using steel flitch

3.3.4 Confinement

Confinement of timber beams using FRP wraps and metal clamps results in an increase in the bearing capacity, shear strength and stiffness [44]. Cracked segments of beams can also be reinforced using this technique. Even though FRP confinement has proved successful in the strengthening of reinforced concrete (RC) columns, there is a host of issues that arise when FRP wrapping solutions are adopted for timber columns of existing structures. Variations in the dimensions of the timber elements (swelling and shrinking) over time resulting from thermo-hygrometric environmental conditions affect the jacketing potential of the FRP wraps. For traditional timber structures of historical and cultural value, such an intervention is unsuitable due to its limited reversibility and impact on the visual appearance of the structures [34].

3.3.5 Post-tensioning

Post-tensioned steel cables positioned using pulleys, spacers, turnbuckles and anchorage systems are used to reinforce floor and roof systems [52, 53] when there is an increase in the magnitude of loading or the existing timber members do not have sufficient bearing capacity in bending. Post-tensioned steel cables limit the deflection of timber beams and rafters, if they are anchored to resisting tensile loads. Appropriate fire-protection systems must be in place if post-tensioning systems are used. In Fig. 12, the excessive deflection of timber beams under bending is prevented by post-tensioned tendons that resist tensile stresses. Different post-tensioning interventions in timber structures are presented in [54].

Fig. 12 Post-tensioned tendons at the Buonconsiglio castle in Trento, Italy

3.4 Reinforcement at the Structural Subsystem Level

The effective seismic reinforcement of the horizontal and vertical subsystems of timber structures, i.e. diaphragm and shear walls, has a significant effect on their global behaviour, provided that the lateral load resistance increases and a continuous load path from the roof to the foundation is ensured. Interventions in timber floors and roofs aim at making them capable of carrying loads in their planes and ensuring that their connections with the walls below remain uncompromised in case of earthquakes [29, 30, 55, 56]. For shear walls, interventions that cause an increase in at least one or more of the following parameters result in improved seismic performance: strength, stiffness, ductility and energy dissipation capacity [48].

3.4.1 Floors

The main aim of reinforcing timber floors in traditional structures is to increase the in-plane as well as out-of-plane stiffness and to improve the bearing capacity [28, 29]. Traditional timber floors consist of a system of beams, joists and decking. Some of the strengthening interventions used for traditional timber floors are shown in Fig. 13 and listed below:

- Overlaying the original timber decking with a topping of concrete, timber or engineered wood (CLT or glulam panels/plywood sheets) [29, 57]: The topping withstands compressive loads while the original timber decking withstands tensile loads. The efficiency of the connections at the interface between the old and new layers is very critical to the success of the intervention. The composite floors can be designed using the gamma method outlined in Annex B of EN 1995-1-1 [58]. The use of timber and engineered wood in the strengthening of wooden floors is

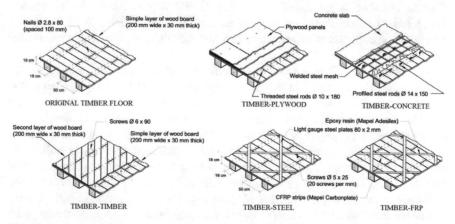

Fig. 13 In-plane strengthening techniques for simply supported timber floors typical of Italian historical buildings [29; adapted] (reprinted with permission from the authors)

preferred as the process is dry and the strength to weight ratio of the new elements is high.

- Overlaying the original timber decking with diagonally placed steel or FRP strips [29]: In the strengthened floor, the diagonal strips carry loads in tension and compression.
- Overlaying the original timber decking with flat steel profiles along the timber joists [59]: In this system, the compressive strength of the timber floor is increased by connecting thin steel plates to the floor joists using dowels. Being dry and reversible, this technique is suitable for timber structures of historic significance.

3.4.2 Roofs

The techniques for in-plane strengthening of timber floors can be applied to timber roofs as well. Additionally, timber roof frames can be reinforced by the addition of steel ties [28]. Since the cross-section of the steel members is small, these members are not visually obtrusive and are low-mass interventions [60]. In roof trusses, the addition of timber bracing elements can be used to counter buckling in members, like the rafters and struts, by reducing the lengths of the members in compression [61]. Interventions in roof trusses may require shoring to relieve the load on the existing truss until the strengthened system is in place. The determination of the jacking loads required and the placement of props is essential to ensuring the safety and stability of the structure [62]. In Fig. 14, the reinforcement of a heavy timber roof truss with supplementary steel members is seen.

3.4.3 Shear Walls

The methods of reinforcing traditional timber shear walls include the addition of diagonal members composed of timber, steel or FRP, application of sheathing material (plywood or OSB), use of mechanical fasteners, steel plates or FRP strips at wall joints, installation of viscous dampers, and the application of reinforced render [48, 63, 64].

Reinforcement techniques that are specific to certain traditional wall types can also be used. For example, in the case of the seismic retrofit of Pombalino walls, experiments have shown that reinforcement solutions employing steel plates (see Fig. 15a) are very effective as they increase the stiffness, lateral load resistance and energy dissipation capacity of the walls [63, 64]. For Pombalino walls without infill or with weak infill, near-surface mounted (NSM) flat steel bars at the wall joints are a viable option as they increase the stiffness and energy dissipation capacity of the walls [65]. The lightweight panels of walls of Ikara and Ekra houses are prone to dislodgement during earthquakes [66]. In Fig. 15b, a typical Ikara wall is seen. Unlike the Pombalino system, there are no diagonal braces to increase the in-plane shear resistance. The use of cross braces in the Ikara system can therefore greatly

Fig. 14 Reinforcement of a
heavy timber roof truss
(reprinted with permissions
from STRUCTURE
magazine, February 2015
[62])

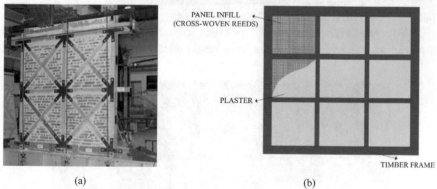

(a)　　　　　　　　　　　(b)

Fig. 15 a Pombalino walls retrofitted with steel plates [63] (reprinted with permission from the
authors); **b** a typical wall of the Ikara system

improve its seismic performance. The use of cross bracing is also advocated in taquezal constructions [17].

3.5 Lessons from Past Interventions

Some of the best lessons in seismic strengthening of traditional timber structures can be learnt from past failures. It is important to discuss and disseminate this information so that such mistakes can be avoided in the future. In this section, a few examples of failed interventions in traditional timber structures are discussed.

In the case of joints, it is imperative to understand that an excessive increase in the strength and stiffness of any of the joint members, resulting from an intervention, might be detrimental to the stability of the structure as it may change the way in which the loads are transferred and induce additional stresses in the members [40]. Therefore, joint interventions need to be carefully designed, especially with regard to rotational capacity and stiffness [67]. An example of over-stiffening of a traditional timber joint is seen in Fig. 16.

Traditional timber floors and roofs are lightweight structures which generate smaller earthquake-induced inertial forces. Therefore, adopting seismic strengthening solutions of significantly higher mass and stiffness at the roof and floor levels can be catastrophic [55, 56, 68]. The inadequate seismic performance of traditional structures in Italy, where strengthening interventions consisted in replacing timber floors and roofs with heavy concrete slabs (see Fig. 17), is a case in point. This led to the Italian building code [69] categorically advocating the preservation of timber roofs and floors (first published in 2005, and then suggested again in 2008 and 2017).

In the case of vernacular timber-laced masonry structures, it is important to bear in mind that certain features might have been added to the system in response to

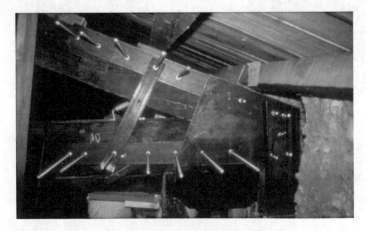

Fig. 16 Intervention on a roof truss that results in excessive stiffening of the joint [32]

Fig. 17 Examples of damage incurred due to the replacement of traditional timber floors and roofs with reinforced concrete elements [68] (reprinted with permission from the authors)

sudden earthquakes. However, the knowledge of their original purpose can get lost over time and they may consequently be regarded as just decorative features or even be eliminated from the structure [70]. This is exemplified in some of the Pombalino buildings of Vila Real de Santo António in the south of Portugal, wherein the frontal walls (see Fig. 18a), which were intended to act as shear walls, have been done away with, thereby increasing the seismic vulnerability of the buildings [16]

(a)   (b)

Fig. 18 a Reconstructed frontal wall without a structural function in a building in Vila Real de Santo António, Portugal (image credits: Javier Ortega) and **b** upper storey of heavy masonry infill walls above lightweight Ekra walls in Sikkim, India [66] (reprinted with permission from the authors)

Figure 18b shows a house in Sikkim, India, which contradicts the design philosophy of progressive reduction of weight from the bottom to the top of a building. The walls of the ground floor are of the lightweight Ekra type, but the upper storey is constructed with heavier brick masonry infill walls. The increase in the mass of the upper floor consequently increases the seismic vulnerability of the structure.

4   Conclusions

For effective seismic reinforcement of traditional timber structures, the following considerations must be made:

- Since moisture problems and biological decay have proven to be responsible for the poor performance of most traditional timber structures in past earthquakes, the adoption of construction details that prevent water ingress into the structure, provide for self-draining of joints and prevent the direct contact or embedment of timber elements in masonry or concrete must be ensured prior to any intervention.
- The key components that contribute to the seismic resistance of different traditional timber systems must be determined. Similarly, the vulnerabilities of these systems must be identified. The structural elements that contribute to seismic resistance, such as ties, ring beams and diagonal braces, must be incorporated in case of their absence.
- Depending on economic and technological feasibility, accessibility and desired performance, seismic reinforcement of a traditional timber structure can be done at the level of the joints, members, or sub-structural systems, using many of the proven methods that have been listed in this report. The improvement in the seismic performance is an outcome of enhancing the strength, stiffness, ductility and/or energy dissipation capacity of the structure.
- The pivotal role played by joints in the seismic response of traditional timber structures requires that they be reinforced with mechanical fasteners, in case of their inability to withstand load reversal without loss of contact. The intervention must not excessively curtail joint movement but should aim at improving the ductility and energy dissipation capacity.
- The efficiency of novel strengthening and retrofitting techniques in terms of strength, compatibility and durability must be checked through appropriate laboratory tests prior to their implementation on site. Their long-term effects on the behaviour of a test structure must be considered prior to their implementation in real structures.

References

1. Feilden B (1987) Between two earthquakes—cultural property in seismic zones. A joint publication of ICCROM and the Getty Conservation Institute
2. Branco JM, Lourenço PB, Aranha CA (2015) Shaking table tests of a two-storey log house. Struct Build Proc ICE 168:803–812. https://doi.org/10.1680/jstbu.14.00073
3. Hirai T, Kimura T, Yanaga K, Sasaki Y, Koizumi A (2004) Lateral resistances of log constructions. In: Proceedings of the 8th world conference on timber engineering (WCTE), Lahti, Finland, vol III, pp 251–254
4. Doğangün A, Tuluk ÖI, Livaoğlu R, Acar R (2006) Traditional wooden buildings and their damages during earthquakes in Turkey. Eng Fail Anal 13:981–996. https://doi.org/10.1016/j.engfailanal.2005.04.011
5. Foliente GC (2000) History of timber construction. In: Kelley SJ, Loferski JR, Salenikovich AJ, Stern EG (eds) Wood structures: a global forum on the treatment, conservation, and repair of cultural heritage, ASTM STP 1351. ASTM
6. Yao K, Zhao HT, Ge HP (2006) Experimental studies on the characteristics of Mortise-tenon joint in historic timber buildings. Eng Mech 23:168–173
7. Tanabashi R (1960) Earthquake resistance of traditional Japanese wooden structures. Bulletins—Disaster Prevention Research Institute, Kyoto University 40:1–15
8. Fintel M (1974) Handbook of concrete engineering. Van Nostrand Reinhold Company Inc., New York
9. BIS. IS:1893 (2007) Indian standard criteria for earthquake resistant design of structures: Part 1 general provisions and buildings, Bureau of Indian Standards, New Delhi
10. Dhandapany D, Arun M (2019) Seismic behavior of timber laced masonry structures in the Himalayan belt. RILEM Book Series, Springer Netherlands. https://doi.org/10.1007/978-3-319-99441-3_65
11. Buchanan AH, Dean JA (1998) Practical design of timber structures to resist earthquakes. In: Proceedings of the 1988 international timber engineering conference, Seattle, pp 813–822
12. Branco JM, Descamps T (2015) Analysis and strengthening of carpentry joints. Const Build Mater 97:34–47. https://doi.org/10.1016/j.conbuildmat.2015.05.089
13. Das R (2007) Standing firm: traditional aseismic architecture in the Western Central Himalayas. In: Agrawal et al (eds) Traditional knowledge system and archaeology, Aryan Books International, New Delhi, pp 49–60
14. Gautam D (2017) Seismic performance of world heritage sites in Kathmandu Valley during Gorkha seismic sequence of April-May 2015. ASCE J Perform Const Facil 31(5). https://doi.org/10.1061/(asce)cf.1943-5509.0001040
15. Vintzileou E (2008) Effect of timber ties on the behavior of historic masonry. J Struct Eng 134(6). https://doi.org/10.1061/(asce)0733-9445(2008)134:6(961)
16. Ortega J, Vasconcelos G, Rodrigues H, Correia M (2015) Local seismic cultures: the use of timber frame structures in the South of Portugal. In: Proceedings of the 2nd international conference on historic earthquake-resistant timber frames in the mediterranean region, H.E.a.R.T
17. Holliday L, Kang THK, Mish KD (2012) Taquezal buildings in Nicaragua and their earthquake performance. J Perform Const Facil 26:644–656. https://doi.org/10.1061/(ASCE)CF.1943-5509.0000266
18. Langenbach R (2015) The earthquake resistant vernacular architecture in the Himalayas. In: Correia, Lourenço V (eds) Seismic retrofitting: learning from vernacular architecture. Taylor and Francis, London, pp 83–92
19. Lopez M, Bommer J, Mendez P (2004) The seismic performance of Bahareque dwellings in El Salvador. In: Proceedings of the 13th world conference on earthquake engineering. Vancouver, Canada
20. Xavier R, Paupério E, Menon A (2015) Traditional construction in high seismic zones: a losing battle? The case of the 2015 Nepal earthquake. In: Correia, Lourenço, Varum (eds) Seismic retrofitting: learning from vernacular architecture. Taylor and Francis, London

21. Chand B, Kaushik H, Das S (2017) Experimental study on traditional Assam-type wooden house for seismic assessment. In: Proceedings of the 16th world conference on earthquake engineering, 16 WCEE. Santiago, Chile, pp 1–11
22. Ortega J, Vasconcelos G, Correia M (2014) An overview of seismic strengthening techniques traditionally applied in vernacular architecture. In: Proceedings of the 9th international masonry conference, Guimarães, Portugal
23. Poletti E, Vasconcelos G (2015) Seismic behaviour of traditional timber frame walls: experimental results on unreinforced walls. Bull Earthq Eng 13:885–916
24. Foliente GC (1998) Design of timber structures subjected to extreme loads. Prog Struct Mat Eng 1(3):236–244
25. Parisi MA, Piazza M (2008) Seismic strengthening of traditional carpentry joints. In: Proceedings of the 14th world conference on earthquake engineering, Beijing, China, pp 1077–1085
26. Palma P, Garcia H, Ferreira J, Appleton J, Cruz H (2012) Behaviour and repair of carpentry connections—rotational behaviour of the rafter and tie beam connection in timber roof structures. J Cultural Heritage 13:S64–S73. https://doi.org/10.1016/j.culher.2012.03.002
27. Branco JM (2008) Influence of the joints stiffness in the monotonic and cyclic behaviour of traditional timber trusses. Assessment of the efficacy of different strengthening techniques. PhD thesis, University of Minho and University of Trento
28. Parisi MA, Piazza M (2015) Seismic strengthening and seismic improvement of timber structures. Constr Build Mater 97:55–66
29. Piazza M, Baldessari C, Tomasi R (2008) The role of in-plane floor stiffness in the seismic behaviour of traditional buildings. In: Proceedings of the 14th world conference on earthquake engineering, Beijing, China
30. Parisi MA, Chesi C, Tardini C, Piazza M (2008) Seismic vulnerability and preservation of timber roof structures. In: D'Ayala F (eds) Structural analysis of historic construction: preserving safety and significance. In: Proceedings of the 6th international conference on structural analysis of historic construction, 2–4 July 2008, Bath, UK. Taylor and Francis Group, London, pp 1253–1260
31. FEMA (1992) NEHRP Handbook of techniques for the seismic rehabilitation of existing buildings. Federal Emergency Management Agency, Washington D.C.
32. Parisi MA, Piazza M (2007) Restoration and strengthening of timber structures: principles, criteria and examples. ASCE 'practice periodical on structural design and construction' 12(4):177–185
33. Pinto L (2008) Inventory of repair and strengthening methods timber. Universitat Politècnica de Catalunya, Escola Tècnica Superior d'Enginyers de Camins, Canals i Ports de Barcelona. Departament d'Enginyeria de la Construcció, 2008 (Advanced Masters in Structural Analysis of Monuments and Historical Constructions (SAHC)), Master's thesis
34. CNR (2007) CNR_DT 201/2005. Guide for the design and construction of externally bonded FRP systems for strengthening existing structures, Timber Structures, National Research Council Advisory Committee on Technical Recommendations for Construction, Rome—CNR
35. Broughton JG, Hutchinson AR, Adhesive systems for structural connections in timber. Int J Adhes Adhes 21, 177–186. https://doi.org/10.1016/s0143-7496(00)00049-x
36. Shiratori T, Komatsu K, Leijten A (2011) Modified traditional Japanese timber joint system with retrofitting abilities. Struct Control Health Monit 15:1036–1056. https://doi.org/10.1002/stc
37. Erman E (2002) Timber joint design: the geometric breakdown method. Build Res Inf 30:446–469. https://doi.org/10.1080/09613210210150991
38. Dietsch P, Brandner R (2015) Self-tapping screws and threaded rods as reinforcement for structural timber elements—a state-of-the-art report. Constr Build Mater 97:78–89. https://doi.org/10.1016/j.conbuildmat.2015.04.028
39. Drdácký M, Urushadze S (2019) Retrofitting of imperfect halved dovetail carpentry joints for increased seismic resistance. Buildings. https://doi.org/10.3390/buildings9020048

40. Parisi MA, Piazza M (2002) Seismic behavior and retrofitting of joints in traditional timber roof structures. Soil Dyn Earthquake Eng 22:1183–1191. https://doi.org/10.1016/S0267-7261(02)00146-X
41. Larsen KE, Marstein N (2000) Conservation of historic timber structures: an ecological approach. Butterworth-Heinemann, Oxford
42. Pizzo B, Schober KU (2008) On site interventions on decayed beam ends. In: Core document of COST action E34, Bonding of timber, University of Natural Resources and Applied Life Sciences, Vienna
43. Franke S, Franke B, Harte A (2015) Failure modes and reinforcement techniques for timber beams—state of the art. Constr Build Mater 97:2–13. https://doi.org/10.1016/j.conbuildmat.2015.06.021
44. Schober KU, Harte AM, Kliger R, Jockwer R, Xu Q, Chen JF (2015) FRP reinforcement of timber structures. Constr Build Mater 97:106–118. https://doi.org/10.1016/j.conbuildmat.2015.06.020
45. Smedley D, Cruz H, Paula R (2008) Quality control on site. In: Core document of COST action E34, Bonding of timber, University of Natural Resources and Applied Life Sciences, Vienna
46. Nowak T, Jasieńko J, Kotwica E, Krzosek S (2016) Strength enhancement of timber beams using steel plates—review and experimental tests. Drewno 59(196):75–90. https://doi.org/10.12841/wood.1644-3985.150.06
47. Yoshinori M, Koichiro Y, Kenji U (1992) A theoretical research on the elastic flexural buckling load of a timber column reinforced with steel plates attached by metal connectors. J Struct Constr Eng, p 436
48. Chang W (2015) Repair and reinforcement of timber columns and shear walls—a review. Constr Build Mater 97:14–24. https://doi.org/10.1016/j.conbuildmat.2015.07.002
49. Tanaka H, Idota H, Ono T (2006) Evaluation of buckling strength of hybrid timber columns reinforced with steel plates and carbon fiber sheets. In: Proceedings of the 9th world conference on timber engineering, Oregon, United States
50. Battle S, Steel T (2001) Conservation and design guidelines for Zanzibar stone town. Aga Khan Trust for Culture, Geneva
51. McCraig I, Ridout B (2012) Practical building conservation. Timber, Ashgate Publishing, Ltd.
52. Corradi M, Osofero A, Borri A (2019) Repair and reinforcement of historic timber structures with stainless steel—a review. Metals Open Access Metall J 9(1). https://doi.org/10.3390/met9010106
53. Jurina L (2006) Interventi di consolidamento delle capriate lignee mediante funi in acciaio (in Italian)
54. Uzielli L (2004) Il manuale del legno strutturale (Handbook of structural timber), vol IV: Interventi sulle strutture, Ed. Mancosu. ISBN-10: 8887017654, 2004 (in Italian)
55. Brignola A, Podestà S, Pampanin S (2008) In-plane stiffness of wooden floor. Proceedings of the 2008 NZSEE Conference, New Zealand
56. Parisi MA, Chesi C, Tardini C (2012) The role of timber roof structures in the seismic response of traditional buildings. In: Proceedings of the 15th world conference on earthquake engineering
57. Gubana A (2010) Experimental tests on timber-to-Cross Lam composite section beams. In: Proceedings of the 11th world conference on timber engineering, WCTE
58. EN 1995-1-1:2004 + A2:2014 (incorporating corrigendum 2006). Eurocode 5: Design of timber structures—part 1-1: general—common rules and rules for buildings. European Committee for standardization, Brussels
59. Gattesco N, Macorini L (2006) Strengthening and stiffening ancient wooden floors with flat steel profiles. In: Lourenço PB, Roca P, Modena C, Agrawal S (eds) Structural analysis of historical constructions. New Delhi
60. Ross P (2002) Appraisal and repair of timber structures. Thomas Telford
61. Ilharco T, Paupério E, Guedes J, Costa A (2010) Sustainable interventions: rehabilitation of old timber structures with traditional materials. In: Proceedings of the SB10Mad sustainable building conference, Madrid

62. Smith N, Vatovec M (2015) Divine design: renovating and preserving historic houses of worship. In: STRUCTURE magazine
63. Gonçalves AM, Ferreira JG, Guerreiro L, Branco F (2012) Seismic retrofitting of Pombalino "frontal" walls. In: Proceedings of the 15th world conference on earthquake engineering, Lisbon, Portugal
64. Poletti E, Vasconcelos G, Branco J (2014) Full-Scale experimental testing of retrofitting techniques in Portuguese "Pombalino" traditional timber frame walls. J Earthquake Eng 18:553–579. https://doi.org/10.1080/13632469.2014.897275
65. Poletti E, Vasconcelos G, Branco J (2015) Application of near surface mounted (NSM) strengthening technique to traditional timber frame walls. Constr Build Mater 76:34–50. https://doi.org/10.1016/j.conbuildmat.2014.11.022
66. Kaushik H, Ravindra Babu KS (2009) Housing report: Assam-type house. In: World housing encyclopedia
67. Parisi MA, Piazza M (2004) Seismic strengthening of traditional timber structures. In: Proceedings of the 13th world conference on earthquake engineering, Vancouver, Canada
68. Borri A, Corradi M (2019) Architectural heritage: a discussion on conservation and safety. Heritage 2(1):631–647. https://doi.org/10.3390/heritage2010041
69. Decreto Ministeriale delle infrastrutture (D.M. 14-01-2008) (2008) Norme Tecniche per le Costruzioni. NTC 2008 (in Italian)
70. Ferrigni F (2005) The recovery of the local seismic culture as preventive action. In: Ancient buildings and earthquakes: reducing the vulnerability of historical built-up environment by recovering the local seismic culture: principles, methods, potentialities, Edipuglia srl

Reinforcement of Traditional Timber Frame Walls

Elisa Poletti, Graça Vasconcelos, and Marco Jorge

Abstract Timber frame walls are common structural elements adopted in many countries for different purposes. They constitute an important cultural heritage of different parts of the world and the necessity often arises to intervene in such structures for their preservation. Different strengthening techniques have been adopted when retrofitting timber frame walls, some traditional and others more innovative. As the response of the walls, particularly to horizontal actions, is governed by their connections, retrofitting is usually concentrated at the joints, but interventions can also be carried out on timber members or on infill. In this chapter, an overview of possible retrofitting techniques is presented, focusing on their advantages and disadvantages and their effects on the overall behaviour of the wall. The presented solutions focus mainly on experimental and in situ interventions performed for seismic purposes.

1 Introduction

Timber frame buildings constitute an important portion of many historical dwellings in the world, constituting a common vernacular architecture with varying characteristics. They became popular for their cheap and easy construction in areas where wood was abundant (North America, Scandinavia, UK), for their good seismic performance (e.g. in Portugal, Italy, Greece, Turkey, Peru), as timber frame walls act as shear walls, as well as for their low weight.

While they are recognized as an important world cultural heritage, many of these buildings have known little or no care during their life, or they have been modified without taking into account the structural response after alterations had been made and without considering concepts such as reversibility and re-treatability.

E. Poletti · G. Vasconcelos (✉) · M. Jorge
Civil Engineering Department, ISISE, University of Minho, Braga, Portugal
e-mail: graca@civil.uminho.pt

© RILEM 2021

J. Branco et al. (eds.), *Reinforcement of Timber Elements in Existing Structures*,
RILEM State-of-the-Art Reports 33,
https://doi.org/10.1007/978-3-030-67794-7_8

This chapter aims to present state-of-the-art traditional and modern strengthening techniques for timber frame walls and discuss their advantages, disadvantages and suitability.

Interventions in timber frame buildings can be necessitated by different problems, e.g. decay as a consequence of poor maintenance, change in use and the consequent need for additional strength, cracks or loosening of the infill materials or the timber joints, and local failures of the timber frame. Many examples [1–3] are available on restoration works done on traditional timber frame buildings, and in some cases the end result is the loss of the original structural system, as some element has been substituted by steel, concrete, or new timber.

Indeed, when intervening on traditional Portuguese half-timbered buildings (the so-called Pombalino buildings, a particular type of timber frame building with external masonry walls linked to an internal timber frame system), a common and extremely invasive practice has been the demolition of the inner part of the building, which is then substituted by reinforced concrete, keeping only the original masonry façades [4] and, therefore, actually losing the original timber frame structure.

Many examples are available on restoration works done on traditional half-timbered buildings [1, 2, 4]. Numerous Pombalino buildings in Lisbon have been retrofitted with fibre-reinforced polymer (FRP) sheets in the connections of the timber frame walls [1], or with damping systems that link the timber frame walls and the outer masonry walls through injected anchors and provide additional bracing [1] (see Fig. 1). Another practice is to project reinforced shotcrete onto the timber frame walls [5], but such a solution could effectively have an overly stiffening effect on the joints. Timber-to-timber interventions are carried out on historic buildings, for example the timber-framed churches in Poland.

(a)

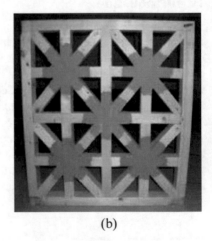

(b)

Fig. 1 Example of strengthening techniques: **a** connections between internal timber-frame walls and external masonry walls; **b** retrofitting with FRP

In the following paragraphs, strengthening solutions will be presented based mainly on experimental results on walls and joints, focusing on strengthening against horizontal actions.

2 Retrofitting of Traditional Timber Frame Walls (Experimental Experience)

In this section, a number of retrofitting techniques that have been studied experimentally are presented. The techniques were studied specifically for protection against seismic actions. During a seismic event, the weakest point of a timber frame wall is its connections, providing that the structural timber surrounding it is sound. Consequently, strengthening interventions are usually carried out on the connections. Additional interventions include interventions on infill, e.g. using reinforced render (see Sect. 3), or the use of damping systems that involve the whole wall and bypass the importance of single joints. When retrofitting timber frame structures, some general principles should be taken into consideration, such as conservation plans (in particular, understanding the structural system of the building and its heritage importance), the causes of deterioration, compatibility of materials, and re-treatability if not reversibility.

2.1 Mechanically Fastened Plates

A traditional method of strengthening timber joints is the use of metallic elements such as steel plates and bars. Steel elements can be screwed, punched or glued.

Steel plates successfully increase the load-carrying capacity of the post-beam connection and are easily implemented. They allow for a better collaboration between the horizontal and the vertical elements and they do not prevent the rotation of the connection. This is not the case when steel plates connect the main post-beam joint to the diagonal bracing. A stiffening effect occurs, thus compromising the ductility of the structure and causing brittle failure in the connections.

An increase in ductility is also observed for walls with weak infill. When using steel plates, there is a minimal loss of the original material and the intervention is potentially reversible. For cultural heritage structures, the possibility of adopting this strengthening solution could depend on its position and visibility and whether it can be hidden by finishing. Cracks may also appear on the plaster due to the presence of steel plates placed above timber. Care has to be taken when applying steel elements at minimum distances from borders, for bolts and screws, and to knots or pre-existing drying fissures, which could create a preferential failure path. In the case of non-machine-worked timber members (non-rectangular section), some difficulty could arise in applying these plates, as contact could not be guaranteed, contrary to what could be obtained when using more malleable strengthening materials, like FRP sheets.

A possible problem that needs to be taken into consideration when using steel plates is the possibility of moisture ingress. To protect such interventions from weathering, stainless or galvanized steel plates should be used.

When timber frame walls are only part of the inner structure (e.g. Pombalino buildings in Portugal), steel plates used to strengthen the walls can be linked to the external masonry walls to prevent the out-of-plane failure of the latter [6]. When the walls have infill, specially crafted plates can be used in order not to cover the infill, which could push and deform the steel plates.

An experimental study carried out on traditional timber frame walls with half-lap joints [7] subjected to quasi-static in-plane cyclic loading adopted such strengthening solution. A steel plate was screwed on either side of the wall at the connection and steel bolts were used to link the two plates of each connection (Fig. 2). The walls had already been tested in the unreinforced condition and retrofitting was applied to the damaged walls, which were appropriately repaired with either prostheses or by the substitution of the element. The results showed that an increase in strength up to 180% could be achieved; even after peak load, a good residual strength was observed for the walls (see Fig. 3). Failure occurred at the joints, but the bolts and steel plates were able to prevent the complete collapse of the connections. Additionally, both the dissipative capacity and stiffness increased after strengthening. For timber frame walls with bracing members (St. Andrews crosses) which originally have weak connections (e.g. nailed), care should be taken not to over-stiffen the connections

Fig. 2 Example of strengthening performed with steel plates secured with bolts

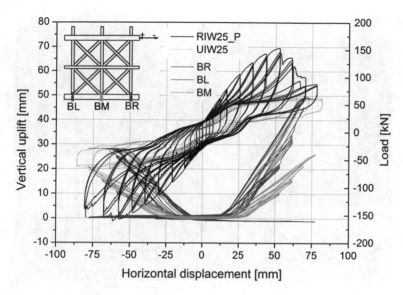

Fig. 3 Experimental results for timber frame walls retrofitted with steel plates (RIW25_P) compared with the unreinforced results (UIW25)

between the bracing members and main frames (see Fig. 4), as this could lead to different structural behaviour of the walls, out-of-plane failure during a seismic event, and a consequent decrease in ductility [7].

The good seismic capacity of steel plates fastened with bolts applied to traditional timber frame walls was confirmed by Gonçalves et al. [5], with over 100% increase in strength and a great improvement in terms of energy dissipation.

Where necessary, steel plates could be used as strengthening for posts, offering a confining effect and preventing buckling.

2.2 Near-Surface Mounted (NSM) Strengthening

A strengthening technique that has acquired popularity in recent years is the application of rods and bars (either in steel or in FRPs) using the near-surface mounted (NSM) method. This technique has proven effective for both flexural and shear reinforcement but it requires specialised workmanship. It can be used as an alternative to externally bonded reinforcement (EBR), albeit it has some advantages over EBR, namely reduced in situ installation work, easier anchorage to prevent debonding, and the possibility of achieving an invisible intervention, since a thin wood cover can be used, though it is not reversible [8]. For a detailed step-by-step illustration of how to perform this retrofitting, see [9]. A possible disadvantage of this technique is that it reduces the timber section, therefore care should be taken not to weaken the effective cross-section.

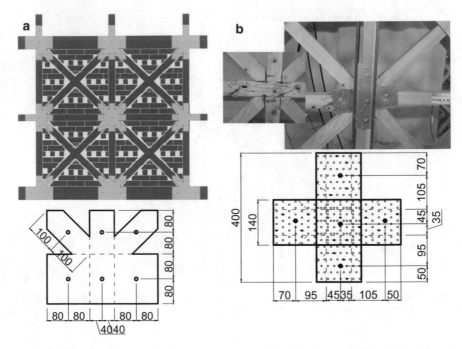

Fig. 4 Retrofitting configuration for timber frame walls: **a** with bracings connected (dimensions in cm); **b** without bracings connected (dimensions in cm)

The following parameters should be taken into account when designing NSM strengthening: (1) cross-sectional dimensions of timber elements involved. Limitations on the minimum distances from the borders should be followed for the cuts; (2) presence of knots or of pre-existing drying fissures. Slots should not be made near knots, since they could weaken the surrounding zones. Important fissures near the intervention zone may need to be filled; (3) tensile strength of bars and rods. Attention should be paid to the type of element used in order to guarantee a sufficient tensile strength to the connection; (4) bond strength between bars and structural glue and between structural glue and the component material. The bond between the materials should be investigated to avoid early failure due to debonding; (5) anchorage length. Moisture ingress should not be considered a problem, as the glue isolates the reinforcement. Additionally, the strength of the glue is guaranteed even if the surrounding timber element is wet. The glues used are specifically designed for timber and are highly compatible with wood.

Eurocode 5 [10] does not provide guidelines for NSM interventions, or any other strengthening intervention, and usually the application of this retrofitting is based on experimental results from the literature (e.g. CNR DT 200 R1/2013, ACI 440.2R-08). NSM strengthening has been applied to timber only in the last two decades. Research has been performed by Jorge [9], who studied the bond behaviour between glulam and FRP and then applied FRP strips with the NSM technique to continuous double span glulam slabs for bending strengthening. From the analysis of the tests, the author suggested that an anchorage length of 15 times the diameter should be used. The same anchorage length is suggested by other authors. For FRP strips, good performances were found for bond length of 7.5 times the height of the strip [11] for concrete structures.

CK45 steel was used for NSM strengthening with steel flat bars, with a tensile strength between 600 and 800 MPa (a value of 672.87 MPa was obtained experimentally). The bars had a section of 8×20 mm^2, so the ultimate tensile force that they could withstand was 108 kN per bar, applied to the half-lap connections of traditional timber frame shear walls, in *Pinus pinaster*, and tested under in-plane cyclic loads [12]. Two bars were welded together to form a cross shape and were inserted in the slots cut in the post and beam of the half-lap joint (Fig. 5). Another half-lap joint was also created between the two bars and then welded to improve the anchorage length. A structural timber glue was used, namely MAPEI Mapewood Paste 140 [13], which is compatible with both materials, has a fast drying time (7 days) and is durable.

An increase in strength of up to 200% was obtained, while maintaining a good deformation capacity. The retrofitting proved to be more appropriate when no or weak infill was present in the walls, as strong infill limits the deformation of timber members and makes it impossible to fully exploit the additional deformation capacity offered by NSM bars. This technique offers a very good shear response to the walls and gives additional strength to the connections, preventing early tensile/shear failure during seismic events and avoiding rocking mechanisms [12]. Stiffness increased by up to 100% and the dissipative capacity by up to 160%, particularly for walls with no or weak infill (see Fig. 6). Moreover, it was observed during the tests that NSM retrofitting hindered the opening of the connections when compared to, for example, a retrofitting performed with steel plates, indicating greater stiffening of the connections. This effect could not always be positive, so great care should be

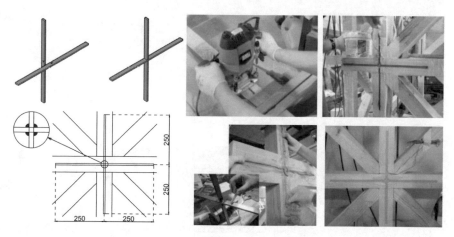

Fig. 5 Example of strengthening performed with NSM flat steel bars (dimensions in cm) [11]

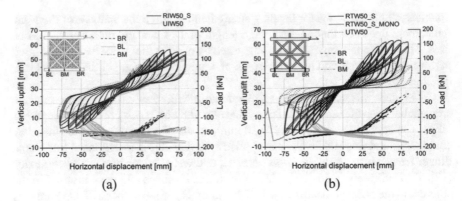

Fig. 6 Experimental results of: **a** infilled wall with NSM strengthening; **b** timber frame wall with no infill

put into the selection and amount of the bar or rod used in terms of strength and deformability of the same.

This retrofitting technique is of easier application than, for example, glued-in rods, because it guarantees easy accessibility and the cuts can be performed directly on the wall in situ when it is not necessary to substitute elements.

Apart from steel elements, NSM strengthening can be performed using fibre reinforced polymers (FRPs), either laminates or rods (see Fig. 7). It was seen, from an experimental campaign carried out on glulam slabs strengthened with NSM rods situated in the tension zone [9], how this type of strengthening could greatly improve the flexural strength of timber beams, thereby increasing their ductility. Additionally, even though it was not designed for the redistribution of any kind of moment, the technique was able to re-distribute the bending moment by approximately 25%.

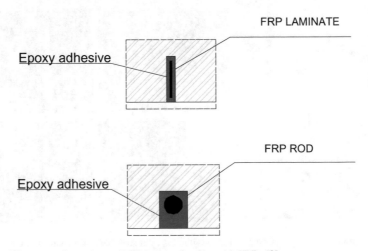

Fig. 7 Different configurations of NSM strengthening with FRPs [8]

Other studies have confirmed the good response offered by this type of strengthening (e.g. [14, 15]) when applied to timber structures. In the case of timber frame walls, if flexural strengthening of the wall beams is deemed necessary, the possible presence of infill and the possibility of its removal in order to proceed with the intervention has to be taken into account.

2.3 Externally Bonded Reinforcement (EBR) Using Fibre-Reinforced Polymers (FRPs)

Fibre-reinforced polymer (FRP) is a composite of fibres and matrix; the fibres provide strength and stiffness and the polymer matrix protects the fibres from abrasion and transfers stresses between them. There has been an increased use of FRPs in timber technology in the last decade because they do not corrode, have a high strength to weight ratio, are non-conductive and non-metallic, and have low maintenance requirements. Different products are available (plates, rods and sheets) and different fibres can be used [16]. However, some drawback exists in terms of durability and long-term performance.

FRPs are frequently used in practice to strengthen existing timber structures (see separate Chapter "Fiber-Reinforced Polymers as Reinforcement for Timber Structural Elements") in this RILEM state-of-the-art report. FRP sheets are glued on wall connections or members to improve their strength, usually based on empirical knowledge, as no standardisation on FRPs exists for timber, only for concrete. National guidelines are available, e.g. CNR-DT 201/2005, concerning the strengthening of timber structures with FRPs. Experimental investigations help to better understand the most appropriate solutions for FRP strengthening.

Cruz et al. [17] performed diagonal tests on reduced scale wallets strengthened with glass fibre-reinforced polymer (GFRP) rods and glass fibre fabric (GFF) sheets. The walls were retrofitted by embedding two GFRP bars to the outer timber members and gluing GFF sheets to the timber elements of the central connections. The strengthened wall panels showed a recovery strength of up to 127% and good improvement in terms of ductility (see Fig. 8).

More research has been carried out on modern timber frame walls, e.g. [18], but for these walls strengthening is not usually considered. An improvement in their seismic capacity is usually achieved through the adoption of different sheathing or through an alternative disposition of the frame [19]. Premrov et al. [20] studied timber frame walls coated with carbon fibre-reinforced polymers (CFRP) strengthened with fibre-plaster boards (Fig. 9). The CFRP strips were applied diagonally and different widths and number of strips were considered. Results showed that, while no increase in terms of stiffness was recorded, the strength of the walls improved.

Poletti et al. [21] performed pull-out and in-plane cyclic tests on half-lap joints retrofitted with GFRP sheets. The tests chosen were meant to capture the hysteretic behaviour and dissipative capacity of the connections and to characterise their

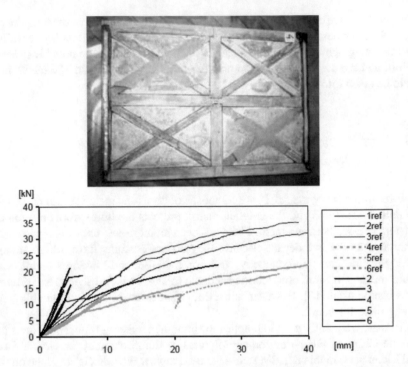

Fig. 8 Wall panels strengthened and tested with FRPs: (top) strengthened panels; (bottom) experimental results (2, 3, 4, 5, 6—initial state of the panels; 1ref, 2ref, 3ref—panels strengthened with FRP rods and GF fabric; 4ref, 5ref, 6ref—panels strengthened with FRP rods only) [14]

Fig. 9 Walls strengthened with CFRP [18]

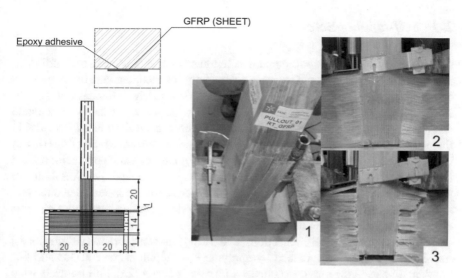

Fig. 10 Example of strengthening performed with GFRP sheets [8, 19]

response and, therefore, their influence on the seismic response of timber frame walls, particularly in regard to their uplifting and rotation capacity, which could lead to rocking in the walls. Uni-directional sheets were applied to both sides of the connection, forming a cross on the connection on the overlapping. The results on pull out cyclic tests showed a very high initial stiffness and the maximum capacity of the connection for a low value of vertical uplift. Failure occurred in two parts, first on one side of the connection with the debonding of the vertical sheet (detail 1 in Fig. 10) and then on the other side with failure of the fibres perpendicular to their direction (details 2 and 3 in Fig. 10). The maximum strength achieved was 15 times greater than that of the unreinforced specimen, but the residual strength was lower than that of the unreinforced specimen. Due to the geometry of the connection and the use of uni-directional sheets instead of multi-directional ones, the fibres also worked perpendicularly to the fibres, leading to their separation and eventual failure. The same problem was encountered during the in-plane cyclic tests, since the failure of the sheets occurred in a direction perpendicular to the fibres at the height where the post met the beam. Nevertheless, the load-carrying capacity increased by 50%.

When considering timber strengthening, particularly of traditional structures, the adoption of CFRP materials is often not cost-effective, since the structure will not be able to mobilise the full capacity of the materials. By using GFRP products, the structure becomes less rigid and higher strains are reached. Carbon based products are able to give better results in terms of creep and fatigue, but it has to be analysed if their additional capacity is effectively needed. Additionally, the visual impact is lower.

2.4 Self-tapping Screws

One of the least invasive reinforcement techniques for timber is the use of additional screws for the connections. The screws are easily inserted, can easily reconnect a cracked element or connection and provide additional strength. Recently, the use of self-tapping screws proved particularly effective when axially loaded (see separate Chapter "Self-tapping Screws as Reinforcement for Structural Timber Elements" in this RILEM state-of-the-art report and e.g. [22]). Inclined arrangement of screws can transfer shear and tensile forces and strengthen bending-resistant connections. Care has to be taken to position the screws at the appropriate angles to exploit their full strength, as well as to follow the minimum distances between fasteners and the end and the edge of the timber to avoid splitting, but otherwise this constitutes an easy, cheap and reversible strengthening technique.

In the case of traditional connections, and taking into consideration seismic loads, half-lap joints were strengthened by applying self-tapping screws at 30° and 60° configurations and then used to connect a post and a beam [23]. Pull-out tests were performed and results showed that this strengthening solution was able to greatly improve the strength of the connection (6 times over) and its stiffness without showing brittle failure; the solution ensured a post-peak softening behaviour and, therefore, a great capacity to dissipate energy. Failure proved to be mild, since the damage increased progressively with pulling out of the screws throughout the test, causing slight damage to the beam (see Fig. 11). After the peak load, the screws were responsible for grain disorganisation. Plastic deformations of the screws were observed at the end of the test (6 mm screws were used). Though in-plane tests were not performed, it is believed that given the inclination of the screws, the solution would have been beneficial and could increase the strength and dissipative capacity of the connection, and consequently of the wall, even if not in such a dramatic way as one of the strengthening solutions mentioned above. From the pull-out tests it is clear that

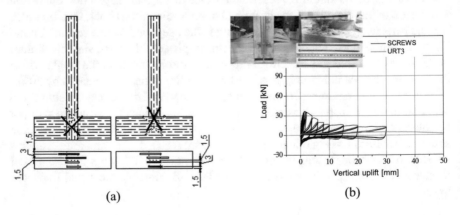

(a)   (b)

Fig. 11 Joint strengthened with self-tapping screws: **a** scheme of screw application; **b** pull-out test results and damage observed

they would help prevent rocking behaviours in timber framed walls under seismic load.

Trautz and Koj [24] strengthened the mitre joints of a rigid timber frame. The strengthened joints were designed for both positive and bending moments. The tests showed a significant higher load-bearing capacity compared to conventional joints with dowels or glued finger-joints.

Tannert and Lam [25] studied the effect of strengthening, by self-tapping screws, of rounded dovetail connections under vertical shear loading, considering different angles between the screw axis and the wood grain of the joist. A significant increase in the capacity and stiffness was observed.

Moisture effects have to be considered since they affect the withdrawal capacity. Metal fasteners act as constraints, and moisture variations can change the strength and stiffness properties of timber, induce cracks and affect the load-carrying capacity of the connections. In that regard, a number of studies have been carried out on the effect of moisture content variation on the withdrawal capacity of self-tapping screws (e.g. [26]).

2.5 Timber-to-Timber

Though the response of timber frame walls is dominated by their connections and strengthening the connections is the most common intervention, failure can still occur in a post or beam, be it due to decay, the presence of a defect, insufficient cross-section for the current load, etc.

Various strategies can be adopted to repair posts and beams or increase their strength. The most traditional repair technique is timber-to-timber intervention [27]. Repair can refer to patching, where only the damaged part of a section is taken out and a new piece is inserted, or the substitution of a whole section; the worst-case scenario may require the replacement of a whole member (Fig. 12). Patch repairs use a combination of glues, bolts, timber dowels and screws to guarantee the continuity

Fig. 12 Timber-to-timber interventions in a church in Jawor, Poland built with traditional timber frames (*credits* E. Poletti)

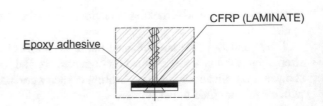

Fig. 13 Mechanically fastened externally bonded plate [8]

of the member and avoid, as much as possible, differential movements between the two parts, which could lead to water trapping. Care should also be taken in selecting the timber for repairs: same species of timber with similar grain orientation and moisture content should be adopted.

Whole section repairs may be necessary before carrying out additional strengthening. A prosthesis is created, in that case, and connected to the existing member. The connection is, once again, the most crucial point. A poorly executed prosthesis could nullify the effects of any other strengthening [23]. Traditionally, different typologies of scarf joints are adopted, with the addition of screws and bolts. More recently, glued-in rods and self-tapping screws are used to secure the prosthesis and existing member.

Other times, it may be necessary to replace the entire timber member, for example when there is extensive decay.

After post and beam repairs have been carried out, the different strengthening solutions presented in this chapter can then be applied to the posts and beams.

2.6 Mixed Interventions

Other types of retrofitting could be adopted for timber frame walls. Though little to no experimental studies are available on it, a strengthening method which proved effective for timber beams and slabs is the externally bonded reinforcement (EBR) with glued plates screwed to the timber element (Fig. 13) [9].

The intervention increased both the strength and ductility of the timber element, but only slightly reduced the timber section.

Similarly, a series of combinations of NSM and EBR techniques can be adopted for posts and beams when appropriate (e.g. NSM and FRP sheets for posts).

An alternative and advanced, but aesthetically invasive, solution is the use of an elasto-plastic steel damper, which consists of steel bars and rods [5] and operates along the diagonal of the timber frame wall in tension, to provide additional dissipative capacity. Results from in-plane cyclic tests showed an increase in strength and energy dissipation of over 100%.

3 Interventions on Infill

Infill plays an important role in half-timbered structures, as it increases both the strength and stiffness of the timber frame. When intervening on infill panels, one has to take into consideration the building's performance and the position of the half-timbered wall. An external unsheltered wall has different requirements than an internal infill wall. It is important not to trap moisture inside the wall and not to alter the connection to the timber frame.

When repairing traditional half-timbered walls, infill is usually replaced with modern bricks and cement-based mortars, which could exacerbate moisture problems by capturing and transferring moisture to the timber frame and joints.

An intervention that affects both the infill and timber frame is the application of reinforced render, which is often performed in practice [5]. It consists in applying a steel mesh covering the whole wall and then spraying it with shotcrete. Tests performed on walls on which reinforced render was applied on both sides [6] showed that this solution greatly increased the stiffness of the walls, without taking advantage of the dissipative capacity of the connections (Fig. 14).

Traditional timber walls used for partition purposes (tabique) but not structural purposes have been retrofitted using earth-based renders to enhance their thermal insulation [28].

4 Discussion

When dealing with the strengthening of timber frame walls, strengthening is usually applied to their connections, as they affect the overall response of the walls [5, 7] since they represent the dissipative mechanism of the wall. All retrofitting techniques discussed in this chapter are able to increase the strength and stiffness of timber frame walls and improve their ductility. Table 1 provides an overview of the strengthening techniques with some recommendations to be considered.

While FRP and NSM retrofitting provide a great improvement to the structure, they are non-reversible, and this could be an issue when dealing with heritage buildings.

Additionally, the durability of all strengthening techniques has to be addressed, particularly for externally bonded solutions, since little information in that regard is available in the literature. Interventions on timber alter its moisture content, and this can affect the whole structure. Moreover, the durability of FRP materials and epoxy resins used in different retrofitting techniques can be affected by ageing and weathering.

The same can be said for interventions with prostheses. In this case, apart from the compatibility of materials and the durability of epoxy resins, it is also important to address the level of continuity between the original structure and the prosthesis and the possibly altered response due to the prosthesis.

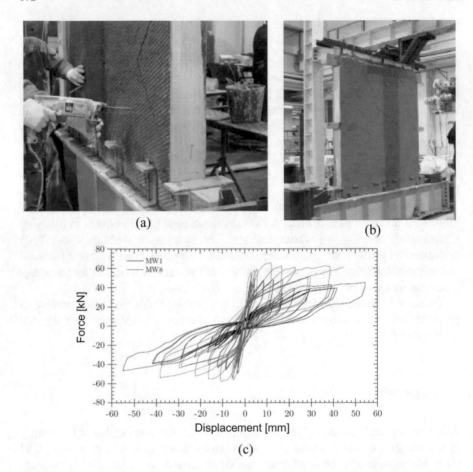

Fig. 14 Timber frame wall with reinforced render [6]: **a** steel mesh applied; **b** final appearance after the application of shotcrete; **c** experimental results (black: unreinforced wall; red: reinforced wall)

5 Conclusions

In this chapter, an overview of different retrofitting techniques for timber frame walls has been given, based mainly on experimental results. Though the focus has been on interventions for seismic actions, these techniques can also be applied to other circumstances due to the adoption of walls such as shear walls. Little research is available on the reinforcement of traditional timber frame walls, but the results available on walls and other elements have shown possibly effective interventions.

Traditional interventions such as timber-to-timber, metal fasteners and steel elements are able to restore and improve the capacity of walls. NSM strengthening, either with steel or FRP elements, is able to greatly increase the capacity and ductility

Table 1 Applicability of strengthening techniques to timber frame walls

Type of strengthening	Compatibility with timber frame walls	Comments
Steel plates	✓	
NSM	✓	Appropriate for connections. For flexural strengthening of other elements, beware of accessibility due to infill
FRP sheets	✓	Appropriate for connections. For flexural strengthening of other elements, beware of accessibility due to infill
Self-tapping screws	✓	
Timber-to-timber	✓	Care in grain direction. When prosthesis is used, is it appropriately connected?

of walls and has the potential of being an invisible intervention. Externally bonded FRPs also gave good results, but their durability has to be addressed.

References

1. Appleton J, Domingos I (2009) Biography of a Pombalino. A case study in downtown Lisbon. Editora ORION, Lisbon (in Portuguese)
2. Tsakanika-Theohari E, Mouzakis H (2010) A post-Byzantine mansion in Athens. Restoration project of the timber structural elements. In: Proceedings of WCTE world conference on timber engineering, June 20–24, 2010 Riva Del Garda, Trento, Italy
3. Aktas YD (2017) Seismic resistance of traditional timber-frame hımış structures in Turkey: a brief overview. Int Wood Prod J 8(1):21–28
4. Cóias V (2007) Structural rehabilitation of old buildings. ARGUMENTUM, GECoPRA, Lisbon (in Portuguese)
5. Gonçalves A, Ferreira J, Guerreiro L, Branco F (2012) Seismic retrofitting of Pombalino "frontal" walls. In: Proceedings of the 15th world conference on earthquake engineering, Lisbon, 24–28 September 2012
6. Moreira S, Ramos LF, Oliveira DV, Lourenço PB (2014) Experimental behavior of masonry wall-to-timber elements connections strengthened with injection anchors. Eng Struct 81:98–109
7. Poletti E, Vasconcelos G, Jorge M (2014) Full-scale experimental testing of retrofitting techniques in Portuguese "Pombalino" traditional timber frame walls. J Earthquake Eng 18:553–579
8. Sena-Cruz J, Branco J, Jorge M, Barros JAO, Silva C, Cunha VMCF (2012) Bond behavior between glulam and GFRP's by pullout tests. Compos Part B 43(3):1045–1055
9. Jorge MAP (2010) Experimental behavior of glulam-FRP systems. MSc thesis, Department of Civil Engineering, University of Minho
10. Eurocode 5 (2004) EN 1995-1-1:2004. Eurocode 5: design of timber structures—part 1–1: general-common rules and rules for buildings, CEN, Brussels

11. De Lorenzis L, Teng JG (2007) Near-surface mounted FRP reinforcement: an emerging technique for strengthening structures. Compos Part B, 38:119–143
12. Poletti E, Vasconcelos G, Jorge M (2015) Application of near surface mounted (NSM) strengthening technique to traditional timber frame walls. Constr Build Mater 76:34–50
13. MAPEI (2002) Mapewood Paste 140. Thixotropic epoxy adhesive for the restoration of timber structural elements (available at: http://www.mapei.com/public/COM/products/1503_GB.pdf)
14. Xu Q, Chen JF, Zhu L, Li X, Zhang F (2012) Strengthening timber beams with near surface mounted carbon fiber reinforced polymer rods. In: Proceedings of the 6th international conference on FRP composites in civil engineering, CICE
15. André A (2006) Fibres for strengthening of timber structures. Research report. Luleå University of Technology:03
16. Schober K-U, Harte AM, Kliger R, Jockwer R, Xu Q, Chen J-F (2015) FRP reinforcement of timber structures. Constr Build Mater 97:106–118
17. Cruz H, Moura J, Saporiti J (2001) The use of FRP in the strengthening of timber-reinforced masonry load-bearing walls. In: Lourenço PB, Roca P (eds) Historical constructions. Guimarães, Portugal
18. Lam F, Prion HG, He M (1997) Lateral resistance of wood shear walls with large sheathing panels. J Struct Eng 123(12):1666–1673
19. Varoglu E, Karacabeyli E, Stiemer S, Ni C (2006) Midply wood shear wall system: concept and performance in static and cyclic testing. J Struct Eng 132(9):1417–1425
20. Premrov M, Dobrila P, Bedenik BS (2004) Analysis of timber-framed walls coated with CFRP strips strengthened fibre-plaster boards. Int J Solids Struct 41:7035–7048
21. Poletti E, Vasconcelos G, Branco JM (2015) Experimental analysis of the cyclic response of traditional timber joints and their influence on the seismic capacity of timber frame structures. In: Proceedings of SHATIS'15, Wroclaw, Poland
22. Dietsch P, Brandner R (2015) Self-tapping screws and threaded rods as reinforcement for structural timber elements—a state-of-the-art report. Constr Build Mater 97:78–89
23. Poletti E (2013) Characterization of the seismic behaviour of traditional timber frame walls. PhD thesis, Department of Civil Engineering, University of Minho
24. Trautz M, Koj C (2009) Self-tapping screws as reinforcement for timber structures. In: Alberto Domingo A, Lazaro C (eds) Proceedings of the international association for shell and spatial structures (IASS) symposium 2009, Valencia
25. Tannert T, Lam F (2009) Self-tapping screws as reinforcement for rounded dovetail connections. Struct Control Health Monit 16:374–384
26. Silva C, Ringhofer A, Branco JM, Lourenço PB, Schickhofer G (2014) Influence of moisture content and gaps on the withdrawal resistance of self-tapping screws in CLT. 9° Congresso Nacional de Mecânica Experimental, Aveiro, Portugal
27. English Heritage (2012) Practical building conservation. Timber. Ashgate Publishing Ltd., Farnham, England
28. Pinto J, Cunha S, Soares N, Soares E, Cunha V, Ferreira D, Briga Sa A (2016) Earth-based render of Tabique walls—an experimental work contribution. Int J Archit Heritage 11(2):185–197

Reinforcement of Light-Frame Wood Structures

Marisa Mulder, Michael Fairhurst, and Thomas Tannert

Abstract Recent post-earthquake evaluations of the performance of light-frame wood buildings have pointed out a number of deficiencies that may make such buildings susceptible to high levels of damage and collapse in a seismic event. This chapter focuses, firstly, on defining the deficiencies that are common in light-frame wood structures, including weak first storey, weak roof or floor diaphragms, shear walls with insufficient strength, inadequate load path, geometric and mass eccentricities, and brittle components. Subsequently, the chapter describes conventional as well as state-of-the-art retrofit solutions that are available in specific codes and guidelines.

1 Introduction

1.1 Background

Light-frame wood structures are the most prevalent construction type in North America, representing over 90% of the residential building stock [1]. Many of these buildings, e.g. over 75% in San Francisco, United States [2] and over 40% in Vancouver, Canada [3], were built prior to the adoption of modern building codes and seismic engineering design practices. Many of these structures were built in a construction era where the use of archaic materials, such as lath and plaster or horizontal boards, and archaic construction practices with little or no detailing for establishing a loading path were applied. As a result, a number of these buildings may be vulnerable during a seismic event due to insufficient strength and stiffness of their seismic force resisting systems, poor load path definition, and vertical/torsional irregularities. The quality of the materials and level of detailing can significantly affect

M. Mulder · M. Fairhurst
The University of British Columbia, Vancouver, Canada

T. Tannert (✉)
University of Northern British Columbia, Prince George, Canada
e-mail: Thomas.Tannert@unbc.ca

© RILEM 2021
J. Branco et al. (eds.), *Reinforcement of Timber Elements in Existing Structures*,
RILEM State-of-the-Art Reports 33,
https://doi.org/10.1007/978-3-030-67794-7_9

the performance and likelihood of collapse during a seismic event [4]. A recent study predicted that up to 80% of the structures would be flagged as unsafe and 25% of existing multi-storey wood buildings would be expected to collapse in a magnitude 7.2 earthquake in the Bay Area of San Francisco [5]. These findings point to a critical need to access and retrofit the existing light-frame wood building stock in all earthquake-prone zones of North America.

1.2 Lateral Force Resisting System in Light-Frame Wood Buildings

Wood shear walls are the main gravity and lateral force resisting system in light-frame wood buildings. The floor and roof diaphragms distribute the gravity and lateral loads to the shear walls. The shear walls then transfer the loads to the next lower level or to the foundation, as shown in the depiction of the load path in Fig. 1 [6]. Wood shear walls, consist of vertical studs, framing members with frame-to-frame connections, sheathing panels and sheathing-to-framing connections. The in-plane lateral resistance is primarily developed through the sheathing-to-framing connections (i.e. nails) in racking action. Furthermore, the connections provide hysteretic damping and energy dissipation under cyclic loading.

1.3 Performance of Light-Frame Wood Structures During Earthquakes

Wood-frame structures have traditionally been considered to perform well, in regard to ensuring the safety of life, during seismic events. This assumption is derived from the inherent light weight and deformation capacity, the structural redundancy and the ability to dissipate energy within the connections. Although this favourable behaviour has been generally observed in many worldwide earthquakes, this is not always the case. There have been several recorded incidences of excessive damage or collapse of light-frame wood structures subjected to significant ground shaking. These cases are usually caused by an easily identifiable structural deficiency, such as a weak first storey, inadequate load path or inadequate anchorage [7].

In the 1971 San Fernando earthquake (magnitude 6.7), many older wooden houses suffered varying levels of damage: from minor damage to structural collapse. But also many newer, multi-storey apartment buildings with large openings at their ground levels were severely damaged. The prominent forms of damage observed were: sliding off foundations, collapse of cripple walls, collapse of non-structural partitions such as porches and chimneys, and collapse or major damage of weak first storeys. The majority of modern (at that time) houses with no major structural deficiencies performed well [8]. Figure 2a illustrates a common type of damage observed

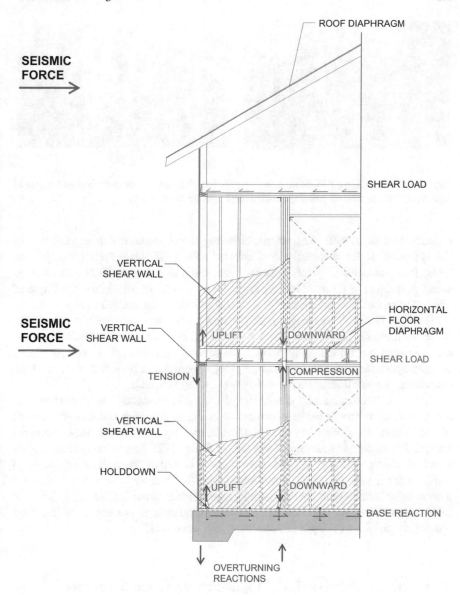

Fig. 1 Light-frame wood building with the load path

in wooden houses in the 1971 San Fernando earthquake: sliding off foundation and collapse of non-structural partitions. Damage caused by weak first storeys is shown in Fig. 2b.

The 1987 Edgecumbe earthquake in New Zealand comprised a magnitude 6.3 main-shock, preceded by a magnitude 5.2 fore-shock and followed by four signif-icant aftershocks (each of magnitude greater than 5.0). The earthquake occurred in

(a)   (b)

Fig. 2 a House sliding off foundation during 1971 San Fernando earthquake; **b** near-collapse of weak first-storey apartment buildings in the 1989 Loma Prieta earthquake

a rural area near several small towns, including Edgecumbe, which was 8 km from the epicentre of the earthquake, and affected nearly 7000 buildings (mostly light, wood-frame structures) [9]. No houses collapsed and less than 50 houses suffered substantial damage; damage was typically due to sliding off foundations, collapse of brick veneer, collapse of brick chimneys, and failure of foundation posts.

The 1989 Loma Prieta (magnitude 7.1) earthquake was one of the most damaging recent earthquakes in Western North America. Although most wood buildings near the epicentre performed well, there were several collapses of older four-storey wooden apartment buildings with large garage openings in their first storeys which caused these weak first storeys to collapse (Fig. 2b) [10, 11].

The 1994 Northridge earthquake (magnitude 6.7) produced the highest ground accelerations ever recorded in an urban area in North America and caused between 30–40 billion U.S. dollars in property damage, making it one of the most expensive natural disasters in the history of the United States [12]. Light-frame wood buildings had undergone structural and non-structural (e.g. gypsum wall board cracking) repairs after a seismic event, where more than $20 billion in losses was directly associated with the repair cost for wood frame residential buildings [13]. In both the 1989 Loma Prieta and 1971 San Fernando earthquakes, a number of multi-storey apartment buildings collapsed due to weak first storeys [12].

2 Common Deficiencies of Light-Frame Wood Structures

2.1 Soft First Storey

A soft- or weak-storey building has one storey with less stiffness or strength compared to the other storeys. ASCE7-10 [14] classifies such vertical irregularity, based on the relative stiffness and strength of the floors, into the following categories: stiffness soft-storey, strength weak-storey, stiffness-extreme soft-storey, and strength-extreme weak-storey; these different categories are shown in Fig. 3. Vertical irregularity is

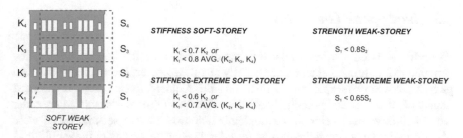

Fig. 3 Soft or weak storey classification in ASCE 7-10 [14]

prevalent in multi-family buildings with parking or retail ground floors that often have large openings for windows or doors and open floor plans with few partition walls [15]. The lateral strength and stiffness of such a ground floor has been observed to be as low as 30–40% of that of the upper storeys [2]. Soft-storey buildings have been recognized as a disaster preparedness problem; these structures are susceptible to pancake collapse during earthquakes. In these buildings the majority of energy dissipation and damage occurs at the soft storey level and the remaining upper floors behave as a rigid body with very limited damage. The soft storey level has been observed in post-earthquake evaluations (e.g. of the Loma Prieta earthquake and the Northridge earthquake) to experience concentrated damage, large residual drift and, in some cases, collapse. Approximately 4000 extreme soft-stiffness storey buildings exist in San Francisco [16], with over 58,000 residents living in these. Due to the large number of these structures there is a need to develop affordable assessment and retrofit solutions to ensure the life-safety of these building in case of a seismic event.

2.2 Weak Diaphragm

For a wood-frame structure, or any other type of structure, to resist the lateral loads imposed by seismic events, it must have an adequate load path. In a light-frame structure the majority of the mass is carried by the diaphragms (floors and roof). When this mass is accelerated, by a strong shaking event for example, it imposes forces on these diaphragms. These forces must be transferred by the diaphragms to the walls of the structure. The walls, which make up the lateral force resisting system of wood-frame structures, must be able to resist the forces and transfer them to the building's foundation to complete the load path. Excessive deformations in inadequate floor or roof diaphragms can substantially increase the severity of earthquake damage. Poor diaphragm performance can lead to significant structural and non-structural damage, although this will typically not result in structural collapse.

2.3 Inadequate Load Path

Wood-frame buildings have relatively small mass, thus the inertial forces they generate due to ground shaking are usually small. However, if any part of the load path is missing or incapable of transferring the required forces, large amounts of deflection, damage, and potentially local or global collapse may result. Many older wood buildings built prior to the implementation of modern seismic codes were designed without adequate load paths for lateral loads, thus many of these buildings present a significant seismic risk. The first seismic regulations were included as an optional appendix in the 1927 *Uniform Building Code* (UBC) and these regulations have evolved greatly since then.

Since light wood-frame structures use diaphragms and shear walls to carry gravity loads, these elements, by themselves, are typically strong enough to resist some lateral load (from a small or moderate shaking event). However, the connections between the elements and to the foundation may not have been designed to adequately transfer forces and retain the load path under lateral loading. Several common areas where existing wood-frame buildings may be deficient are: connections to diaphragms, connections through floors, and anchorage to the foundation. Figure 4 illustrates the lateral load path of a light-frame wood structure with an adequate load path on the left and a poor load path on the right. Each "link" of the load path is numbered on

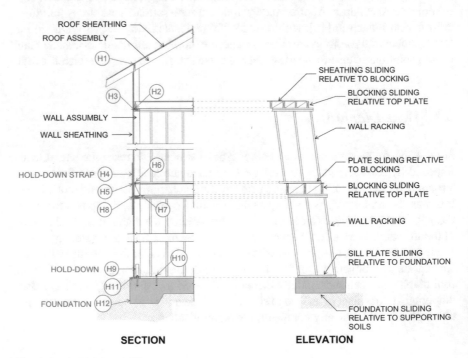

Fig. 4 Lateral load path [6]

the left: from H1 to H12; the consequence of each missing link is illustrated on the right [6].

2.4 Inadequate Connections

In multi-storey platform-frame buildings, the shear walls are not continuous over the height of the structure. The connections through floor assemblies between storeys must have adequate strength to transfer shear forces between walls from different storeys. The inadequacy or failure of these connections can lead to large amounts of damage in the upper storeys. Diaphragms must be properly connected to their supporting shear walls and must be detailed with adequate drag struts/chords in order to transmit forces through the them and into the supporting walls. If any of these components are inadequate, excessive diaphragm deformations may occur under significant lateral loads. Poor diaphragm performance can lead to large amounts of damage to the structure.

Most wood frame structures constructed prior to the 1970s do not have adequate anchorage and overturning resistance in case of a seismic event. Provisions for anchorage did not exist until the late 1950s and these requirements were not widely adopted until the late 1970s [17]. In post-earthquake evaluations, failures in the sill plate and structures shifting off the foundation have been observed [18]. Fragility analysis of conventional residential constructions in the Central US concluded that additional nailing and hold-downs could significantly improve the expected performance and reduce the probability of overturning failures [17]. Shear walls with adequate anchorage perform well in racking. Racking deformation of shear walls yields high hysteretic damping and energy dissipation under seismic loading. When a wall does not have proper anchorage, such as hold-downs, the failure of the wall is governed by other modes of deformation, including rocking and sliding of the walls. These modes are, in most cases, associated with reduction in global strength, stiffness and the energy dissipation capacity [18]. Shear wall tests indicated that there was large reduction in strength (averaging a 66% reduction) and failure displacement (averaging a 35% reduction) of walls without hold-downs compared to walls with hold-downs [19]. Figure 5 illustrates the behaviour of a wood building with a) sufficient overturning restraints (hold-downs and strapping) and b) insufficient overturning restraint and the consequent unrestrained overturning deformation. The "links" in the overturning load path are numbered from OV1 to OV8 on the left. The consequence of any missing link in the overturning load path would be wall uplift, as illustrated on the right.

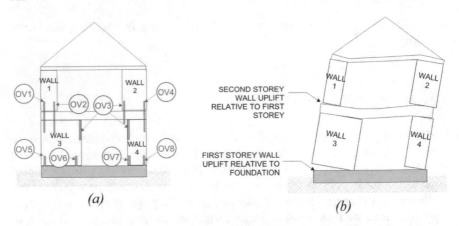

Fig. 5 Comparison of the overturning load path connections and deformation of wood building **a** with and **b** without adequate overturning resistance [6]

2.5 Geometric and Mass Eccentricities

Torsional irregularity occurs when the center-of-mass (CM) and center-of-rigidity (CR) of a storey do not coincide. This can happen in buildings with irregular distribution of mass, irregular geometry of the floor plans, or irregular distribution of lateral force resisting elements. Ground floors with open store fronts or garages commonly have irregular distribution of the lateral force resisting elements. Eccentricity between the CM and CR can cause torsional sensitivity and in-plane torsional moments and can lead to, stress concentration, severe damage and, in some cases, collapse.

2.6 Brittle Components

Under seismic loading, ductile components, with the ability to deform after yielding, perform much better than brittle components. Ductile elements have significant inelastic deformation capacity and therefore can dissipate substantial seismic energy. Because of this property, these elements can typically resist higher levels of shaking without major structural damage or loss of their lateral load capacity. Nailed shear walls with wood structural panels, which are used in modern light wood-frame structures, are typically quite ductile; however, some existing buildings can do with sheathing materials with much lower displacement capacities. Ceccotti and Kara-cabeyli [20] showed that adding a more brittle material (gypsum wall board) to a ductile system (OSB wall board) could limit the ductility of the whole system, which may lead to more damage, or even collapse, in a significant shaking event. Brittle sheathing materials are also commonly used in upper storeys, where the seismic demand is less critical, which may lead to damage or premature collapse of these storeys. Such sheathing materials include stucco or plaster on wood or gypsum lathe.

3   Standard Retrofit Techniques

3.1   Retrofit of Shear Walls

One simple and efficient method of increasing the lateral resistance of an existing light-frame wood structure is by strengthening its existing shear walls. This can usually be done with minimal disruption to the building; because it utilizes the existing components, the floor plan of the building can remain unchanged. To increase the capacity of the existing shear walls, extra nailing can be added to the existing panels and frames; however, more typically, the walls will need to be re-sheathed. If an existing wall is unblocked, then new solid blocking will need to be installed at all sheathing edges. In addition, if the existing foundation connections are inadequate, hold-downs and, possibly, new anchor bolts should be installed. In some cases, a new grade beam may need to be installed below the shear walls if the existing beam is insufficient for the higher loads that would be transferred from the stronger, retrofitted shear walls. This will increase the cost of the retrofit and will require much more work and time.

3.2   Retrofit of Diaphragms

Many older wood buildings have floor and/or roof diaphragms sheathed with shiplap or tongue and groove decking which may not provide enough capacity to resist seismically induced forces. A typical retrofit in this situation would be to re-sheath the diaphragm with new plywood, as presented in Fig. 6. Flat metal straps (drag struts/chords) must also be added along the diaphragm perimeter and any drag lines. This will ensure forces are "collected" from the diaphragm and redistributed to the shear walls.

3.3   Retrofit of Load Path

The main lateral force resisting elements in light wood-frame structures are the diaphragms and shear walls. Not only do these elements have to be strong and stiff enough to resist earthquake damage, the connections between them and the anchorage to the foundation must be strong enough to provide a proper lateral load path. Inadequate connections or missing load paths can severely increase the amount of damage during a strong shaking event. This is typically a problem for older wood-frame buildings that were designed to out-of-date seismic code provisions or designed without any seismic standard at all. See Fig. 4 for an illustration of a typical load path for a light wood-frame structure.

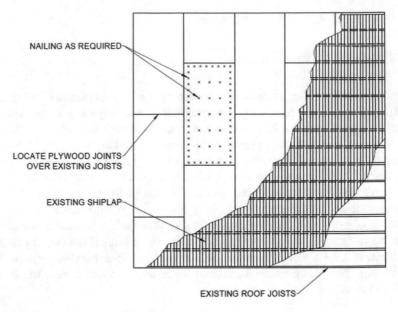

NAILING AS REQUIRED

LOCATE PLYWOOD JOINTS
OVER EXISTING JOISTS

EXISTING SHIPLAP

EXISTING ROOF JOISTS

Fig. 6 Example retrofit of a shiplap diaphragm [6]

Connections Through Floors

In many buildings where the shear capacity of the existing shear walls needs to be increased through additional nailing, additional plywood, or other measures, the existing connections through floors may not be sufficient to transfer this new shear demand to the walls below. In this case, new blocking may be installed between the upper and lower shear walls and connected with framing angles or by nailing through a new bottom plate to allow the transfer between storeys of larger forces imposed from the strengthened walls.

Connections to Diaphragms

Existing buildings, with poor connections of plywood diaphragms to the top of the shear walls or with inadequate shear capacity of the diaphragms, will require the connections to the diaphragms to be upgraded. For these buildings, a typical retrofit would comprise installing new blocking between existing joists, nailing the existing (or upgraded) plywood diaphragms to the joist blocking, and then connecting the existing (or upgraded) shear walls to the blocking through new framing angles.

Anchorage

As seen in many large, historic earthquakes, inadequate foundation anchorage can be a major cause of damage [7]. When existing shear walls are upgraded through increased nailing, increased edge blocking, or re-sheathing, the existing anchorage may not be adequate to transfer the new, larger shear forces imposed by the stronger walls to the foundation. New anchor bolts epoxied to the existing foundation are typically appropriate for this retrofit, if the existing foundation has enough capacity. If the foundation also requires retrofit, then this method can also be used to anchor the walls to the new foundation.

3.4   Retrofit of Structures with Eccentricities

A structure with geometric and mass eccentricities is typically very sensitive to torsional effects and is prone to excessive drifts and damage during significant seismic events [21]. Therefore, significant geometric or mass eccentricities are typically not allowed in retrofitted structures. A straight-forward remedy to this type of deficiency is the addition of new lateral force resisting components to offset the eccentricity.

3.5   Retrofit of Brittle Components

One methodology of retrofitting a structure with brittle components is to only retrofit the first storey, but keep the first storey resistance low enough that damage does not spread to the upper storeys [22]. One benefit of this approach is that the construction is limited to the first storey, allowing the rest of the building to remain operational during the retrofitting. Because this approach is simple and efficient in terms of construction time and cost, it may be adopted by jurisdictions with a large number of deficient structures that need remediation. Another approach is to replace the brittle walls with new, more ductile components. This allows for more efficient and optimized retrofit solutions, as damage can spread over all the storeys of the structure, rather than concentrate on a single storey. This methodology has been shown in full-scale testing to provide an efficient energy dissipation mechanism, allowing structures to resist larger shaking events, compared to a first-storey only retrofit [4, 23]. This approach, however, will likely increase the time and complexity of the retrofit and will put the structure out of operation for much longer.

4   Assessment and Retrofit Guidelines and Solutions

4.1   FEMA P-807

The FEMA P-807 (*Seismic Evaluation and Retrofit of Multi-Unit Wood-Frame Buildings with Weak First Stories*) document proposes assessment and retrofit methodologies for multi-storey wood frame buildings with soft first storeys. The document proposes affordable retrofit solutions to ensure an acceptable level of performance, such as the life-safety during the design earthquake and shelter-in-place following the design earthquake [15]. The methodology focuses on retrofitting only the "soft" first storey to reduce the costs of the retrofit and ensure that the upper storeys of the building can remain in service during the rehabilitation. The document gives guidance on the recommended lower and upper bounds of the strength and stiffness for the retrofit, increasing the displacement capacity of the soft storey and reducing torsional sensitivity where possible. The relative strength method is applied to optimize the benefits of the ground storey retrofit. The idea is to ensure that the ground floor is not over-strengthened such that the soft-storey behaviour moves up to the higher floor levels. The methodology explicitly considers the strength of all components, including the non-structural walls. It is important to note that ignoring these elements is not necessarily conservative. The guidelines are also developed with the objective of ensuring an inexpensive implementation of the design. This means that the majority of the document is prescriptive in nature. To apply the retrofit solution, details regarding the floor plan, diaphragm assemblies, building materials and condition, connection hardware and location, and seismic weight need to be obtained either on-site or from structural drawings.

A retrofit solution using FEMA P-807 must meet the criteria on: (i) eligibility constraints, (ii) strength requirements, and (iii) an eccentricity limit. Several retrofit solutions are recommended in the FEMA P-807 document. Wood shear walls can be strengthened with hold-downs, additional blocking and re-sheathing. Steel retrofit solutions such as moment-resisting frame systems, cantilever column systems, and buckling-restrained braced frames are also recommended. Retrofit solutions that include concrete or masonry walls, or steel braced frames with very high relative stiffness, are not suitable [15]. It should be noted that detailing for the load path is not explicitly covered in FEMA P-807. The analytical procedures in FEMA P-807 assume that the diaphragms behave rigidly and have adequate strength to transfer the shear forces. The diaphragm may need to be strengthened and the retrofit elements may need to be placed such that they reduce the span of the diaphragm. The retrofit elements must also be placed in such a way as to minimize torsion and optimize the performance. The load transfer, anchorage, and foundations should also be investigated and retrofitted if the components are deficient for the retrofitted performance level [4].

4.2   FEMA P-50 and FEMA P-50-1

FEMA P-50 (*Simplified Seismic Assessment and Retrofit Guidelines for Detached, Single-Family, Wood-Frame Dwellings*) provides a simple seismic assessment form that was developed for national application in the U.S. for single-family, light-frame wood homes [22]. The assessment form enables an inspector to quickly assign a "Seismic Performance Grade" to a structure and to identify deficiencies in the dwelling that may require remediation. The assessment is based on: (1) foundation systems; (2) superstructure framing and configuration; (3) general condition assessment; (4) non-structural elements, age, and size; and (5) local site conditions. The form identifies and applies penalties to common structural deficiencies such as inadequate foundation anchorage and geometric eccentricities to determine a Seismic Performance Grade. The Seismic Performance Grade is used to predict the seismic performance of the structure and to determine the extent of any retrofit. FEMA P-50-1 includes more guidance for retrofitting a potentially hazardous dwelling based on its identified deficiencies [24]. These FEMA documents were designed to be simple and prescriptive, allowing the rapid assessment, rating and retrofit of a large number of structures.

4.3 International Existing Building Code

The *International Existing Building Code* (IEBC) [25] provides requirements on safety and stability for existing and historic buildings. The code encourages the use of existing buildings and covers topics such as repairs, alterations, additions, and changes of occupancy. The IEBC is similar to codes for new constructions, but allows for reduced design forces, recognizing the fact that it may not be realistic or economically feasible to bring an existing or historic building up to new building standards. The IEBC also offers prescriptive provisions for remediating common deficiencies in existing buildings.

Chapter A3 of the IEBC, titled "Prescriptive Provisions for Seismic Strengthening of Cripple Walls and Sill Plate Anchorage of Light, Wood-Frame Residential Buildings", offers prescriptive provisions for light, wood-frame buildings with structural weaknesses involving the foundations, foundation sill plate anchorage, and cripple wall bracing. Measures to ensure quality control are also provided. Chapter A4 of the IEBC, titled "Earthquake Risk Reduction in Wood-Frame Residential Buildings with Soft, Weak or Open Front Walls", covers the analysis and design of retrofit elements for light, wood-frame structures with soft or weak first storeys. This chapter is structurally similar to building codes for new structures; the focus, however, is on reduced design forces, first storey effects, and applications to existing structures. This chapter has been adopted for several seismic mitigation programmes in Californian jurisdictions including San Francisco, Oakland, Berkeley, and Alameda [15].

4.4 ASCE/SEI 41

ASCE/SEI 41, *Seismic Evaluation and Retrofit of Existing Buildings* [26], is a comprehensive and well-recognized document that provides measures to assess the performance of existing buildings and guidance on retrofit measures. ASCE 41 offers three levels (or "tiers") of analysis, from simple screening (Tier 1) to a sophisticated systematic evaluation (Tier 3) involving nonlinear dynamic response history analysis. A "deficiency-based" evaluation and retrofit (Tier 2) is also included, which

provides measures to identify and remediate potential deficiencies in existing buildings. Vertical irregularities, lack of diaphragm or shear wall strength/stiffness, and
foundation anchorage are some of the typical deficiencies identified for wood-frame
structures by these guidelines. One of the major downfalls of ASCE/SEI 41 is that,
due to its complexity and use of nonlinear dynamic response history analysis, it
requires very specialized knowledge and is difficult to implement.

4.5 Seismic Retrofit Guidelines (SRG)

The *Seismic Retrofit Guidelines* (SRG) [27] are an innovative set of performance-
based guidelines for risk assessment and retrofit of British Columbia (BC) school
buildings. The SRG guidelines use incremental nonlinear time history analysis to
determine minimum base shear requirements for existing buildings as an alternative
to a code-based approach. The key philosophy of the approach is to achieve life safety
by limiting the probability of collapse to acceptable levels instead of concentrating on
damage prevention. The SRG Guidelines also provide a library of well-established
and/or previously implemented retrofit solutions for common structural deficiencies.
Diaphragm, existing shear wall, and connection retrofit examples are all included for
existing light, wood-frame buildings.

5 State-of-the-Art Retrofit Solutions

5.1 Retrofit Testing Program

Several testing programmes aim to establish and verify retrofit techniques for light-
frame wood structures. One example is a full-scale four-storey wood-frame building
with a soft and weak first storey that was subjected to a series of seismic shake table
tests as part of the NEES-Soft project [23]. The building was selected such that it is
similar to a typical San Francisco Bay Area soft-storey wood frame structure. The top
three storeys were each designed as two two-bedroom apartment units; the bottom
storey was designed as a parking garage with several large openings.

The high wall density in the upper storeys combined with the large openings in
the first storey created a very soft and weak first storey. The building represented a
corner building with two neighbouring buildings to its north and west. Therefore, the
north and west first storey walls had no openings and were much stiffer than the south
and east walls. This configuration created a large irregularity in geometric stiffness
in the already vulnerable first storey. The test structure was subjected to the 1989
Loma Prieta earthquake and the 1992 Cape Mendocino earthquake, scaled from 0.2
to 1.8 g of the maximum considered earthquake (MCE) [23].

The building was retrofitted and tested in multiple phases using two retrofit methodologies: soft-storey retrofit only (as described in the FEMA P-807 Guidelines) and performance-based seismic design (PBSD). PBSD involves modelling the entire structure and distributing the stiffness and strength throughout the height of the structure so that all floors would be damaged and contribute to the dissipation of the seismic energy [4]. With the FEMA P-807 retrofits, the majority of the damage and deformation was concentrated on the first storey—very little damage was transferred to the upper storeys. In the PBSD retrofits, damage was distributed over the height of the structure, and this helped it resist higher intensities of ground shaking.

The NEES-Soft project involved researching the feasibility of retrofitting with steel special moment frames (SSMF) and inverted moment frames (IMF), rocking cross-laminated timber (CLT) walls, energy dissipation systems (dampers), distributed knee-brace (DKB) systems, and shape memory alloy devices. The retrofit solutions and testing have been described as they pertain to the NEES-Soft project, as well as to additional research. These tests demonstrated that retrofit solutions were able to adequately meet the performance objectives defined by the two retrofit methodologies.

5.2 Steel Special Moment Frame (SSMF) and Inverted Moment Frames (IMF)

Special moment frames (SMF) and inverted moment frames (IMF) are viable retrofit options for light-frame wood structures. Full-scale testing and analysis of the systems indicate that the retrofit is able to meet the performance requirements regarding the strength, ductility and relative stiffness [28, 4]. Pinned-ended SMFs, such as the Strong-Frame SMF system, as shown in Fig. 10, were designed to be suitable as a retrofit solution. These frames have minimal interference with garage openings and other architectural details. The beam-to-column connections are designed such that the plastic hinge occurs away from the column and eliminates the potential for lateral torsional buckling of the beam. The diaphragm flexibility makes the design and installation of the bracing for the beam challenging. The SMF is easily assembled on-site. It has bolted connections that do not require specific training to install. There are no welded connections, in which case the cost associated with certified welders, field inspection and fire risk is reduced. Shear forces from the first floor diaphragm can be transferred to the foundation by connecting the beam with nails to the floor diaphragm. Finally, the base connection is pinned, therefore no moment is produced at the column-to-foundation connection and the foundation would need to be retrofitted to resist the vertical and shear forces [4, 29].

As part of the NEES-Soft project, the FEMA P-807 methodology was implemented to design and retrofit the structure with a single steel SMF in each orthogonal direction in the first storey only. The frames were placed to reduce torsion as much as possible without interfering with the garage parking space. The first-storey SMF retrofit was capable of meeting FEMA P-807 requirements at a shaking intensity of 1.1 g.

The structure was also retrofitted using a design based on the performance-based seismic retrofit (PBSR) developed as part of the NEES-Soft project [28]. In the PBSR methodology, the strength and stiffness throughout the height of the building can be modified to optimize the performance of the building according to the displacement-based design originally proposed by Priestley [30]. Two SMFs were installed in each orthogonal direction in the first storey and additional structural panels were added in all four storeys to increase the strength and stiffness of the structure over its entire height. The retrofit was designed for the maximum considered earthquake shaking level, which had a 1.8 g 5% damped spectral acceleration at the period of the structure (0.35 s).

5.3 Cross-Laminated Timber

Recent research has focused on establishing that Cross-Laminated Timber (CLT) rocking walls are a viable retrofit solution for light-frame wood structures [31]. As part of the NEES-Soft project, a CLT rocking wall retrofit was tested numerically and experimentally [28]. The CLT rocking wall was confirmed suitable for the FEMA-P807 Guidelines. The CLT wall retrofit met the performance criteria by providing adequate strength to the soft storey (4% drift limit at an acceleration of 1.14 g) and ensuring that the damage was not shifted to the upper storeys. The CLT panels were designed to rock and behave primarily in rigid-body motion. Vertically slotted holes at the top shear transfer connection were installed to allow for free rocking. The primary energy dissipation of the walls occurred in the mechanical connections, brackets and hold-downs [32, 33]. The 16-mm diameter threaded rods at each end of the CLT walls were designed to resist the overturning moment and yield for ductility. A metal connector and 6.5-mm diameter self-tapping wood screws were used as shear connectors between the CLT panel and the base steel (i.e. the foundation).

5.4 Energy Dissipation Systems (Dampers)

Energy dissipation systems or dampers have also been proposed as a retrofit solution in wood frame structures. The advantages of using a fluid viscous damper (FVD) are they offer a supplement mechanism of dissipation seismic energy and reduce the forces on the shear walls. FVDs can be installed on the soft-storey of a building. The dampers are often velocity dependent and should be out-of-phase with the displacement-dependent forces from the shear walls [34]. This property would prevent the transfer of large forces to the upper levels, implying that the damper system might be able to achieve compliance with the P-807 relative stiffness methodology [35]. The use of dampers as a seismic protection system for light-framed wood

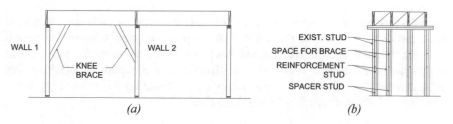

Fig. 7 DKB system: **a** testing of system; **b** elevation view of details [40]

structures was studied [34–39] testing full-scale soft storey building with an energy dissipation retrofit and showed that the energy dissipation system was able to meet the performance objectives. Further work on FVDs to develop them as a practical and reliable method is required.

5.5 Distributed Knee-Brace System

The Distributed Knee-Braced (DKB) system was tested as a possible retrofit solution for the NEES-Soft project for performance-based design and the FEMA-P807 Guidelines. Each individual knee-brace frame is constructed using an additional stud connected to the existing stud, a Simpson Strong-Tie© A35 connector between this stud and bottom plate, a Simpson Strong-Tie© H2A connector between the stud and joist, and two new diagonal 2 × 4 wood members between the reinforced stud and joist fastened with 8d framing nails, as shown in Fig. 7. The knee-brace connections to the studs and joists were designed to act as the system fuse and protect the other framing members and connections. Individual knee-braced systems were installed on a number of frames on the ground floor so that the existing walls and floor members did not contribute to the lateral resistance and the foundation demands were reduced [40]. Reversed-cyclic testing, numerical modelling, hybrid testing, and shake table testing were used to validate the performance of the DKB system. The system was found to provide sufficient strength at very high drift levels and had the potential to be a viable retrofit solution with further development and research [40].

5.6 Shape Memory Alloy (SMA) Devices

Research on using shape memory alloy devices as a retrofit option for light frame wood structures has also been explored [41–44]. The SMA device can dissipate energy and has re-centring capabilities. The device can be built into a compact scissor brace system (34 in. length) that can be installed on an existing structure. This makes it a suitable retrofit solution for storeys with short wall segments and large openings (i.e. garage doors). The SMA systems can also be installed in such a way as to minimize torsion by minimizing the eccentricity between the building's centre of mass and rigidity. Several numerical studies and pseudo-dynamic hybrid (numerical and experimental) tests have confirmed that SMA devices show promise as a retrofit technique and can provide sufficient strength and energy dissipation at the life-safety performance level [23, 45]. Further research on developing this technology into a practical and reliable retrofit solution is required.

6 Conclusions

Existing light-frame wood structures, due to the extremely large percentage of the residential building stock they comprise, present a big risk to cities and communities in the event of a devastating earthquake, especially in North America. This is coupled with the fact that many of these buildings were designed and built before the adoption of modern seismic building codes. Data from previous major worldwide and North American earthquakes suggest that light-frame wood structures, particularly those with structural deficiencies, may be at risk of significant damage or even collapse under significant ground excitation [7]. Several jurisdictions in California, including San Francisco, Oakland, Berkeley and Alameda [15], as well as the province of British Columbia in Canada [27], have implemented large-scale seismic mitigation programmes to address the deficiencies in this type of constructions. This chapter has summarized some common structural deficiencies of light-frame wood structures observed after previous earthquakes and introduced a number of typical retrofit solutions. The main deficiencies included are soft (weak) first storey, weak roof or floor diaphragms, shear walls with insufficient strength, inadequate load path, geometric and mass eccentricities, and brittle components. A section describing state-of-the-art retrofit solutions that are currently under development and/or testing has also been included. Finally, common standards that include provisions for the retrofit of light-frame wood structures have been highlighted, including FEMA P-807 [15]; FEMA P-50 and P-50-1 [22, 24]; The *International Existing Building Code* [25]; ASCE/SEI 41 [26]; and The British Columbia *Seismic Retrofit Guidelines* [27].

References

1. CUREe (1998) CUREe-caltech woodframe project newsletter [Online]. Available: http://www.curee.org/projects/woodframe/
2. Scawthorn C, Kornfield L, Seligson H, Rojahn C (2006) Estimated losses from scenario earthquakes affecting San Francisco: CAPSS—part 2. In: 8th US national conference on earthquake engineering, San Francisco, CA
3. Ventura C, Finn WL, Onur T, Blanquera A, Rezai M (2005) Regional seismic risk in British Columbia—classification of buildings and development of damage probability functions. Can J Civil Eng 32(2):372–387

4. Bahmani P (2015) Performance-based seismic retrofit (PBSR) methodology for multi-story buildings with full-scale experimental validation. Colorado State University
5. ATC (2008) Here today—here tomorrow, earthquake safety for soft-story buildings. Applied technology council, Redwood City, CA
6. FEMA (2006) Homebuilders' guide to earthquake-resistant design and construction. National Institute of Building Sciences, Washington, DC
7. Rainer J, Karacabeyli E (2000) Wood-frame construction in past earthquakes. In: Proceedings 12th world conference on earthquake engineering, Auckland, New Zealand
8. Pacific Fire Rating Bureau (1971) San Fernando earthquake
9. Pender M, Robertson T (1987) Edgecumbe earthquake: reconnaissance report. Bull. New Zealand society of earthquake engineering
10. Bruneau M (1990) Preliminary report of structural damage from the Loma Prieta (San Francisco) earthquake of 1989 and pertinence to Canadian structural engineering practice. Can J Civ Eng 17(2):198–208
11. Harris SK, Egan JA (1992) Effects of ground conditions on the damage to four-story corner partment buildings. In: The Loma Prieta, California, Earthquake of October 17, 1989 strong ground motion and ground failure—Marina District, Washington, D.C., United States Geological Survey, pp 181–194
12. EERI (1996) In: Holmes W, Somers P (eds) Northridge earthquake of January 17, 1994 reconnaissance report, vol 2. Earthquake Engineering Research Institute, Oakland, CA
13. Pei S (2007) Loss analysis and loss based seismic design for woodframe structures. Colorado State University
14. ASCE (2010) American Society of Civil Engineers. In: Minimum design loads for buildings and other structures, ASCE 7, Reston, Virginia
15. FEMA (2012) Federal emergency management agency. Seismic evaluation and retrofit of multi-unit wood-frame buildings with weak first stories. FEMA P-807, Washington, DC
16. ATC (2010) A community action plan for seismic safety. Applied Technology Council, Redwood City, CA
17. Pang W, Rosowsky DV, Ellingwood BR, Wang Y (2009) Seismic fragility analysis and retrofit of conventional residential wood frame structures in the Central United States. J Struct Eng 135(3):262–271
18. Christovasilis I (2010) Numerical and experimental investigations of the seismic response of light-frame wood structures, University of Buffalo
19. Toothman AJ (2003) Monotonic and cyclic performance of light-frame shear walls with various sheathing materials, Virginia Polytechnic Institute
20. Ceccotti A, Karacabeyli E (2002) Validation of seismic design parameters for wood-frame shearwall systems. Can J Civil Eng 29(3):484–498
21. Bebamzadeh A, Ventura CV, Fairhurst M (2015) Plan eccentricity sensitivity study for performance based seismic assessment and retrofit. In: The 11th Canadian conference on earthquake engineering, Victoria, BC, Canada
22. FEMA (2012) Simplified seismic assessment of detached, single-family, wood-frame dwellings, FEMA P-50, Federal Emergency Management Agency, Washington, DC
23. van de Lindt J, Bahmani P, Pryor SE, Mochizuki G, Gershfeld M, Pang W, Ziaei E, Jennings E, Symans M, Shao X, Tian J, Rammer D (2014) Overview of the NEES-soft experimental program for seismic risk reduction of soft-story woodframe buildings. In: Proceedings of the structures congress 2014, pp 2875–2885
24. FEMA (2012) Simplified seismic retrofit guidelines for detached, single-family, wood-frame dwellings. FEMA P-50-1, Federal Emergency Management Agency, Washington, DC
25. ICC (2012) International existing building code. International Code Council, Washington, DC
26. ASCE (2013) Seismic evaluation of existing buildings, ASCE/SEI 41–13. American Society of Civil Engineers, Reston, VA
27. APEGBC (2013) Seismic retrofit guidelines, 2nd edn. Association of Professional Engineers and Geoscientists, Burnaby, BC, Canada

28. Bahmani P, van de Lindt JW, Gershfeld M, Mochizuki GL, Pryor SE, Rammer D (2014) Experimental seismic behavior of a full-scale four-story soft-story woodframe building with retrofits I: building design, retrofit methodology, and numerical validation. J Struct Eng E4014003. https://doi.org/10.1061/(asce)st.1943-541x.0001207
29. Pryor SE, Murray TM (2013) Next generation partial strength steel moment frames for seismic resistance. Research, development, and practice in structural engineering and construction. In: Proceedings of the First Australasia and South-East Asia structural engineering and construction conference, Perth, Australia, pp 27–32
30. Priestley M (1998) Displacement-based approaches to rational limit states design of new structures. In: 11th European conference on earthquake engineering, Paris, France
31. Karacabeyli E, Douglas B (2013) CLT handbook: cross-laminated timber. FPInnovations, Vancouver, BC, Canada
32. Popovski M, Schneider J, Schweinsteiger M (2010) Lateral load resistance of cross laminated wood panels. In: Proceedings of the world conference on timber engineering, Riva del Garda, Italy, 2010
33. Pei S, van de Lindt J, Popovski M (2013) Approximate R-factor for cross-laminated timber walls in multistory buildings. J Archit Eng 19(4):245–255
34. Symans M, WF C, Fridley K (2002) Base isolation and supplemental damping systems for seismic protection of wood structures: literature review. Earthquake Spectra 18(3):549–572
35. Tian J (2014) Performance-based seismic retrofit of soft-story woodframe buildings using energy-dissipation systems. Rensselaer Polytechnic Institute
36. Symans M, Cofer W, Du Y, Fridley K (2004) Seismic behavior of wood-framed structures with viscous fluid dampers. Earthquake Spectra 20(2):451–482
37. Christovasilis I, Filiatrault A, Wanitkorkul A (2007) Benchmark test seismic testing of a full-scale two-story light-frame wood building: NEESWood. NEESWood Project Report
38. Shinde J, Symans M (2010) Integration of seismic protection systems in performance-based seismic design of woodframed structures. NEESWood Project Report
39. Tian J, Symans M, Gershfeld M, Bahmani P, van de Lindt J (2014) Seismic performance of a full-scale soft-story woodframed building with energy dissipation retrofit. In: 10th national conference in earthquake engineering, Anchorage, AK
40. Gershfeld M, Chadwell C, van de Lindt J, Pang W, Ziaei E, Amini M, Gordon S, Jennings E (2014) Distributed knee-braced (DKB) system as a complete or supplemental retrofit of soft-story wood-frame buildings. In: Proceedings of the Structures Congress 2014, pp 1437–1447
41. Jennings E, van de Lindt J (2012) Numerical retrofit study of light-frame wood buildings using shape memory alloy devices as seismic response modification devices. J Struct Eng
42. Jennings E, v. d. Lindt J, JW X, Pang W, Ziaei E (2014) Full-scale hybrid testing of a soft-story woodframe building seismically retrofitted using shape memory alloy devices in Scissor-Jack. In: 10th national conference in earthquake engineering, Anchorage, AK
43. van de Lindt J, Potts A (2008) Shake table testing of a superelastic shape memory alloy response modification device in a wood shearwall. J Struct Eng
44. Jennings E, van de Lindt J, Ziaei E, Mochizuki G, Pang W, Shao X (2014) Retrofit of a soft-story woodframe building using SMA devices with full-scale hybrid test verification. Eng Struct 80:469–485
45. Pang W, Ziaei E, Shao X, Jennings E, van de Lindt J, Gershfeld M, Symans M (2014) A three-dimension model for slow hybrid testing of retrofits for soft-story wood-frame buildings. In: 10th U.S. national conference on earthquake engineering, Anchorage, AK
46. Yanev P, Thompson CT (2009) Peace of mind in earthquake country: how to save your home, business, and life

Reinforcement of Historic Timber Roofs

Eleftheria Tsakanika and Jorge M. Branco

Abstract This chapter aims to present a comprehensive report on the reinforcement of historic timber roofs, focusing on their main characteristics, advantages and disadvantages, which would help professionals select and define the design of reinforcement solutions. Cultural heritage issues are taken into consideration. Reinforcement can be done via different methods—traditional and modern—using simple or sophisticated techniques. An overview of the main materials and the techniques used for selected case studies are presented, illustrating how various reinforcement methods are implemented in practice.

1 Introduction

Historic timber roofs constitute an important part of the cultural heritage of many countries of the world. The increased sensitivity towards the preservation of cultural heritage has led to the adoption of restoration techniques that comply with generally accepted conservation principles and guarantee, as much as possible, the preservation of authenticity and integrity of the structure, minimal interventions, reversibility and compatibility with the original parts of the timber [1, 2, 3]. The aim and scope of intervening in historic structures are dictated by economic, environmental, historical and social reasons.

For historic buildings, repair and/or reinforcement is preferred to total structural replacement, since the authentic materials and structural systems, the construction technology and workmanship, constitute an important cultural value of the historic buildings which needs to be protected and preserved, even if they may no longer be visible, hidden by plasters or ceilings, after the restoration [1, 4]. This growing sensibility towards the preservation and maintenance of heritage buildings, the various

E. Tsakanika
School of Architecture, National Technical University of Athens, Athens, Greece

J. M. Branco (✉)
Civil Engineering Department, ISISE, University of Minho, Braga, Portugal
e-mail: jbranco@civil.uminho.pt

© RILEM 2021   195
J. Branco et al. (eds.), *Reinforcement of Timber Elements in Existing Structures*,
RILEM State-of-the-Art Reports 33,
https://doi.org/10.1007/978-3-030-67794-7_10

species of wood and the complexity of their structural behaviour, their degradation caused by different agents and the need for rehabilitation to incorporate new uses, has led researchers to study different repair and reinforcement solutions.

In most cases all over the world, traditional buildings involve timber used at least as floors and roof systems. The technical and technological development of roof carpentry was very different from country to country. A huge amount of different types of roof structural systems and joints exist, which bears testimony to the diversity and richness of the timber cultural heritage and, at the same time, to the difficulty of studying, repairing and reinforcing them.

Reinforcement or strengthening deals with interventions that increase and upgrade the original or existing load-bearing capacity of the structures, while maintenance and repair try to recover the original load-bearing capacity and return the existing fabric to a known earlier state. Reinforcement is usually applied to extend the use of structures approaching the end of their design life and to ensure that recent requirements for a new use of a building ("heavier" loads or level of safety) and changes in regulations are fulfilled. The reinforcement of the timber parts of masonry buildings, especially their roofs and floors, can prevent also exceptional damaging incidents and improve significantly the overall seismic resistance of the buildings.

Each reinforcement solution has its advantages and disadvantages in regard to conservation philosophy, architecture, aesthetics, structural performance, and technological and construction quality. Economic issues such as the cost of the intervention and the availability of specialized staff can also determine the choice of the method used [5]. When the reinforcement of a roof is being designed, all of the above has to be taken into account and evaluated carefully to ensure that the proper intervention is chosen. A careful choice of the strengthening materials and the reinforcing method is necessary. No material or method can be considered the optimal one. Each case is unique.

This chapter will focus on the reinforcement of historic timber roofs (members and the overall roof behaviour), but not on the reinforcement of existing roofs, which constitutes a much wider group [6]. It also focuses on reinforcement methods for the main load-bearing elements, but not for the secondary ones (purlins, decking) or non-structural elements such as roofing or ceiling materials (clay tiles, timber shingles, ceiling planks, etc.).

In several cases, roofs have important decorative (woodcarving and polychrome) details, markings, symbols and finishes, and very often they carry simple ceilings or ceilings with very high artistic value. In many of these cases, it is required to operate on the spot without dismantling the carpentry or any of its parts, increasing the difficulty both of the assessment and the restoration procedure.[1]

---

[1]Interventions on historic timber structures have different conservation philosophy depending on the cultural background of each country. For example, the logic inherent in Japanese timber buildings and the necessity of ongoing repair by dismantling (approximately every 150–300 years), especially for roofs, call for a different approach to that of the preservation of timber buildings in other regions of the world. This kind of approach is highly appreciated, too, since it is dictated by a deep and long cultural tradition that has existed for centuries [1, 7]. Moreover, the partial or even total dismantling of the structure could be accepted if: (a) repairs carried out in situ and on the original elements

2 Causes of Damage to Timber Roofs (Pathology)

The first step before any intervention (repair or reinforcement), is the documentation and the assessment of the existing timber structure: the understanding of the structural system, the damage and the causes, and the residual strength and stiffness properties. Briefly, the most common and major problems (*pathology*) of timber roofs are: (i) decay problems, usually in parts that water enters and accumulate, such as in support areas (timber parts embedded in the external walls); (ii) insect attack (active or not); (iii) damage or lack of strength and/or stiffness of single members (failures, shrinkage cracks, excessive deformations, etc.); (iv) damage or lack of strength and/or stiffness of joints (failures, shrinkage cracks, etc.); (v) lack of stiffness of the whole timber roof (in-plane or out-of-plane, vertical or horizontal deformations).

Damage and failures of timber roofs can be due to different causes: (i) natural defects of wood; (ii) biological degradation (rot, insect attack); (iii) environmental and atmospheric agents (changes in wood moisture content); (iv) fire; (v) errors in the original conception/poor original design (lack of adequate sections, poor quality of timber, errors in the original structural system); (vi) poor execution; (vii) excessive loading (wind, earthquake, etc.); and, (viii) maintenance or intervention errors during their life time.

Proper assessment with appropriate techniques is obviously of major importance and, therefore, the study of relevant state-of-the-art reports, the scientific work and publications concerned with the diagnostic procedures, is highly recommended [6, 8, 9].

3 Reinforcement Methods

In this chapter, different examples of reinforcement methods will be presented according to the following categories: (i) reinforcement of timber roof members (rafters, tie-beams, posts, end-beams, etc.); (ii) reinforcement of the overall load-bearing system of the roof (improvement of the overall stability, e.g. bracing).

3.1 Reinforcement of Timber Roof Members

To increase the flexural strength and stiffness of timber members (beams), reinforcing elements are usually added to supplement the existing elements. A large variety of reinforcement configurations are available. The reinforcing elements can be in the form of rods, plates, straps or other structural shapes, which are connected to the

---

would require an unacceptable degree of intervention, (b) the distortion of the structure is such that it is not possible to restore its proper structural behaviour, and (c) inappropriate additional work would be required to maintain it in its deformed state [1].

beam using mechanical fasteners or structural adhesives. These reinforcing elements can be placed inside or outside the member and may be passive or pre-stressed. Apart from the structural requirements, the strengthening configuration selected for a particular application may depend on other factors too: aesthetics can limit the use of different materials; the presence of decorative ceilings or painting on beams may require that the reinforcement be restricted to the top or the sides of the timber elements; fire protection, aesthetical issues and other requirements may exclude the use of externally bonded plates on exposed surfaces; geometrical, architectural or constructional limitations can restrict the use of new elements or elements with certain dimensions; etc. [10, 11].

3.1.1 Pre-stressing Metal Reinforcement

Pre-stressing has emerged as one of the most common reinforcement techniques for increasing the bending load-carrying capacity and the stiffness of timber members, when large deflections are observed. It is mainly used for rafters and tie-beams, and may be required due to an inadequate section or low strength and stiffness properties. An important advantage of this reinforcement is its reversibility.

In many publications since the second half of the nineteenth century, it is emphasized how the empirical work of many engineers has created a broad selection of layouts and structural solutions that work properly for many decades (Figs. 1 and 2). Technical manuals refer to this kind of reinforcement, mainly pointing out the difficulties of installing the outer tendons at the head of the beams due to the fact that the beams are embedded into the walls [12, 13]. Of course, there have been

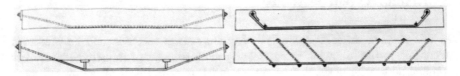

Fig. 1 Different methods for strengthening existing timber beams using pre-stressing techniques [12, 13]

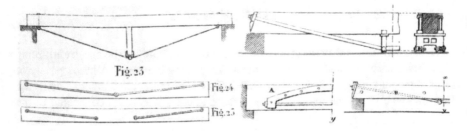

Fig. 2 Reinforced beams with metal tendons by A. R. Emy [12, 13]

Fig. 3 Reinforcement of a truss rafter in the theatre of Sarteano (left) and in a Renaissance palace in Rome by G. Tampone (right) [14]

improvements in these techniques since then, boosting confidence in the use of this kind of reinforcement.

It was not until the 2nd half of the twentieth century that a systematic approach to post-tensioning restoration methods of timber structures was established. Some examples were reported in manuals written during that period and are still a reference for present professionals. Most of the main restorers belong to the Italian school, which is known as a very active centre for restoration theories and projects (Fig. 3).

Nowadays, for the reinforcing system, high strength materials with minimum dimensions such as steel (stainless or not), titanium or carbon fibres can be used.

This system of reinforcement can be applied to sound, unbroken beams of regular shape or to members that have already been repaired [15].

One major issue that needs special attention is how the rheological behaviour of the material and the shrinkage of the timber sections may affect the loss of pre-stressing, especially in cases where the application of the load is transferred by compression perpendicular or at an angle to the grain. If possible, tension perpendicular to the grain should be avoided.

The effect of the environmental and the material's initial and existing conditions, even the conditions after the restoration works, must be taken into account too [16]. Periodic inspections are necessary for the pre-stressing methods, in order to ensure the efficiency of the intervention according to the design specifications.

3.1.2 Non-pre-stressed Metal Reinforcement

A similar concept to the above, but without pre-stressing, can be used in the reinforcement of tie-beams that present excessive deformation. Steel elements may support the transfer of loads by the tie-beam to other elements (members or joints), which are carefully chosen and verified structurally (Figs. 4, 5, 6 and 7) to ensure the safe transfer of the forces through the new load path.

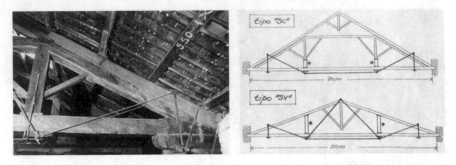

Fig. 4 Reinforcement of a truss tie-beam. Detail and design schemes of the Savona Theatre project (courtesy of the designer, Ing. L. Paolini)

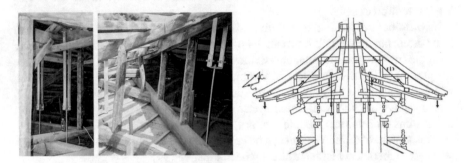

Fig. 5 Post and beam timber roof of Hagi Mehmet Aga Mosque in Rhodes, Greece. Improvement of the original load-bearing system which caused high bending moments and deformation on the tie-beam, by suspending the tie-beam with steel rods from the central upper joint of the roof (left) [17]. It is worth mentioning that the Japanese have a relatively successful history of steel reinforcements of timber structures dating back to the latter half of the eighteenth century. During the repair in 1986 of the three-storied pagoda of the Kiyomizu-dera Buddhist temple in Kyoto, Japan, seventeenth century (right), turnbuckles were introduced to reinforce the roof structure, carrying the load of the deformed projecting eaves [7]

3.1.3 Connection of New Elements (Timber or Steel) to the Existing Timber Members by Steel Fasteners

This is a common technique of reinforcement used to increase the load-carrying capacity of a timber element (e.g. a rafter or a tie-beam), or if the deflection of the beams is too high.

Steel sections and plates (Figs. 9, 11), solid timber sections, or wood-based products (glued-laminated timber, plywood, cross-laminated timber, laminated veneer lumber, etc.), nailed, screwed or bolted either to the tensile face or the vertical sides of the timber beams, are used to repair or reinforce timber elements (Figs. 8, 9 and 10). Similar techniques are used for the substitution of the decayed parts of timber members (see Sect. 3.1.8).

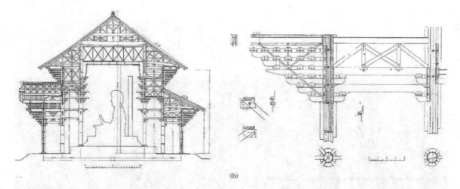

(b)

Fig. 6 The Great Buddha Hall (Daibutsuden) of the Todai-ji temple, Nara, Japan, the world's largest historic timber structure, was given an overall steel reinforcement during repairs in 1903-13. A double steel roof-truss was inserted to support the frame of the large roof, all bracket complexes were reinforced with flat steel bars, rafters and extended beams were reinforced with thick steel plates, and the central part of pillars was reinforced with steel channels [7]

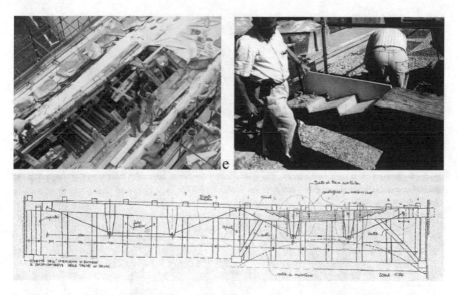

Fig. 7 San Giovanni Battista Church in Salbertrand, sixteenth century: phases of the intervention; scheme of the reinforcement intervention with Polonceau system [18]

As timber beams generally fail in tension in a brittle way, the positioning of the reinforcement at the tensile face of the beams is very effective for increasing bending strength. The above interventions, in most cases, are not applicable to exposed timber structures. Besides its poor aesthetic appeal, reinforcement with external metal plates may suffer other disadvantages due to condensation on the timber members and their consequent vulnerability to decay, dimensional changes of the steel parts caused by

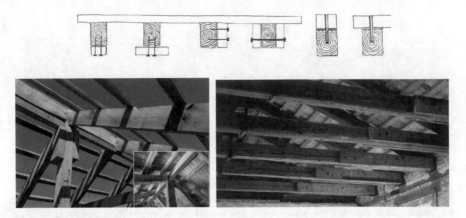

Fig. 8 New timber pieces can be connected with metal fasteners (nails, screws, bolts, dowels, glued-in-rods) to the original timber beams in various positions to improve their load-carrying capacity and stiffness [19, 20]

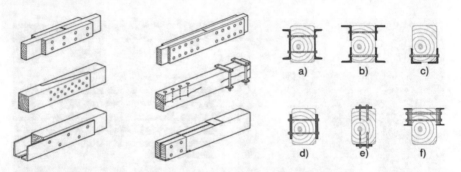

Fig. 9 Repair or reinforcement of timber members using new timber elements or steel plates and sections connected to the original timbers by metal fasteners [5, 21]

Fig. 10 Repair of a timber rafter using large pieces of wood screwed at the sides of the rafter [22]

Fig. 11 Strengthening of a timber member by insertion of steel plates. Invention patent G. Tampone, L. Campa, 1987 (left) [15]. Barn (fourteenth century) in Herefordshire UK (Sinclair Johnston, 2009). Truss repair using a 20-mm thick steel flitch plate and resin. If resins were not used, many bolts would have been required which would have further cut away the fabric and had less aesthetic appeal (center and right) [16, 24]

changes in temperature and, therefore, additional internal stresses in the wood and fasteners, and dimensional changes of the timber that the steel parts cannot adapt to if humidity changes occur. Galvanized steel or protected from corrosion sections, plates or fasteners would be preferable [23].

Similar techniques are used for the substitution of the decayed parts of timber elements (see Sect. 3.1.8).

Internal elements such as steel plates inserted in the timber members can be used too (Fig. 11). They can be connected to the timber beams with screws, steel dowels or bolts.

The system can be applied to undersized, overloaded or broken timber members. It requires geometric regularity of the section of the timber member. For a well-sized but broken element, the length of the inserted plate can be limited to the length of the affected section plus an additional length (one and a half the depth of the member), including sound wood at both sides [15]. Strengthening of timbers with steel flitch plates and resin, working as a composite member, is a method used in several restoration projects in England (Fig. 11, right).

3.1.4 Connection of Timber-Based Sheathing Materials or Panels to the Existing Timber Beams by Metal Fasteners

Some reinforcement techniques commonly adopted in practice for floor beams can be used for roofs too. For example, wooden panels or slabs can be placed over and/or under the existing beams (floor beams, rafters or tie-beams that usually carry ceiling loads). Different configurations are possible depending on the connection system (nails, screws, dowels, rods) and the sheathing material (additional timber planks over the existing ones, plywood, OSB, LVL and recently, slim panels of CLT) (Fig. 12). In this way, the section of the beam is increased (composite T section or H section), thereby increasing the load-bearing capacity in bending of the timber members, minimizing their deflections, and upgrading their in-plane stiffness (vertical plane). The new composite section also ensures a significant upgrade in out-of-plane stiffness

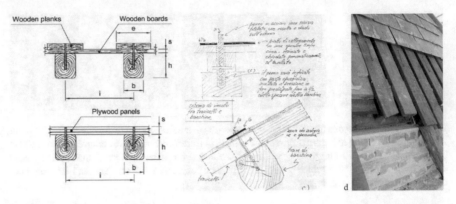

Fig. 12 Examples of reinforcement techniques of timber members, using either new sheathing elements (left) [26], or by improving the connection of the original planks to the main load-bearing elements [18, 27]

(horizontal or inclined), improving the diaphragmatic action of the roof and consequently, if the roof is well-connected to the walls, the lateral load resistance of the buildings (in particular of masonry ones) in seismic areas [25].

In timber-to-timber composite sections, the use of traditional materials and dry assembly methods are in agreement with the restoration demands of compatibility and reversibility of the intervention. An advantage of this method is that the additional loads are quite small due to the use of wood-based sheathing materials (panels, slabs or planks). For all composite sections, the mechanical characteristics of the connections are the main factor which affects the structural response. Design of composite sections requires the consideration of partial composite action, due to the impossibility of achieving an extremely rigid shear connection between the web and flange (deformable shear connection between web and flange). Analysis can follow the Eurocode 5 approximate 'gamma method' [28], where an effective flexural stiffness $(EI)_{ef}$ for the composite section is calculated as a function of the stiffness of the shear connection, taking into account the slip between the flange and the joist. Also, the 'shear analogy method', where the composite beam is divided into two virtual components coupled with stiff bars, can be used to determine the internal forces [13]. It is interesting to note that taking into account the contribution of the original planks of a roof in the structural assessment (e.g. structural models), especially if they are nailed directly on the rafters, that heavy and unnecessary interventions can be avoided by just improving their connection [18].

3.1.5 Addition of New Timbers (Struts or Posts) to Increase the Supports of the Existing Timber Members

This is a typical method of reinforcement for "*post and beam*" roofs, which is the most common type of roofing for buildings in the Balkan and Minor Asia area during

Fig. 13 *"Post and beam"* roofs of the Ottoman period mansions in Greece. Excessive deformation of the longitudinal horizontal beams that support the rafters (left). Reinforcement by adding new timber posts and struts to increase the supports of the longitudinal beams that carry the rafters (right) [25]

the Byzantine and Ottoman periods [29]. It is a spatial system which functions in a completely different way from the well-known types of king or queen post trusses widely used in Italy and other European countries.

The loads are transferred from the rafters through a three-dimensional system of beams, posts and struts not only to the outer walls, but mainly to the internal ones (Fig. 13) [29].

In the *"post and beam"* system, the connections of the posts and struts are capable of transferring compression forces but not tension. Typical damage of such roofs is the bending failure or excessive deflection of the longitudinal horizontal beams that support the rafters (Fig. 13, left), due to the absence of adequate struts or lack of proper sections. The reinforcement of the original load-bearing system can be accomplished by the addition of new struts (denser supporting) (Fig. 13, right), which is an easy, low-cost and reversible intervention, that gives the possibility to keep the original beam in position and retain the concept of the original structural system [17, 20].

3.1.6 Addition of New Timber or Steel Members Next to the Original Timber Members of the Roof

In some cases, new timber or steel elements can be placed parallel to the original ones without any connection, in order to diminish the loads that the existing elements carry (Fig. 14). The concept of this reinforcement is similar to the one described in Sect. 3.2 regarding the whole roof structure.

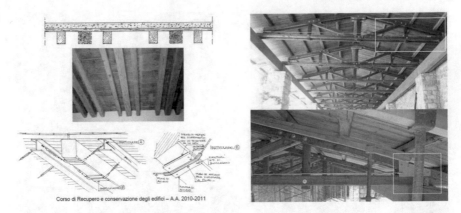

Fig. 14 New timber or steel elements placed between the original beams in order to carry part of the loads (left) [19]. Queen's Tower Estate Ilion, Athens: New steel tie-rod added next to the original timber tie-beam without any connection with it, in order to transfer the tensile loads (right)

Fig. 15 New timber beams used to replace only the severely damaged rafters (left, middle) [25, 20]. The Neoclassical School in the Medieval city of Rhodes: New timber glued laminated beam replaced the poorly constructed original horizontal beam (right)

3.1.7   Replacement of the Whole Timber Member

There are cases (poor initial construction, severe damage or failure, etc.) where the replacement of the whole timber member (rafter, tie-beam, post, strut, etc.) is the only solution. But this should be resorted to, only as a last alternative in any restoration project (Fig. 15).

3.1.8   Beam-End Repair and Reinforcement (Prosthesis)

Intervention in roofs very often involves end-beam repairs. End-beams embedded in masonry walls are the most exposed zones to biological agents. If a high level of moisture is present in the masonry, for example due to infiltration when damage occurs in the roofing elements (e.g. tiles), and the adsorbed moisture cannot be evaporated because of poor ventilation, the suitable conditions for biotic attacks (moisture contents above 20%) are established and degradation of end-beams can, therefore, be expected. Nowadays, several techniques and methods can be found in the literature for the repair and reinforcement of rotten timber end-beams. Since the early 1970s, many companies have developed materials and techniques to repair timber elements by partial substitution of the decayed part (design of a *prosthesis*). All of these techniques aim to restore the load-bearing capacity of the original member and the structural continuity of the old part with the new, thereby ensuring their connection and collaboration.

The intervention consists in the substitution of the decayed part by a new element which can be made of solid wood or wood-based products (glued-laminated timber or LVL) (Figs. 16, 19, 20, 21, 22, 23, 24 and 25), steel sections or plates (Fig. 16b, c), or epoxy resin (Fig. 16d). Such new elements are connected to the original timber part by steel fasteners (nails, screws, bolts, dowels, metal straps, etc.), by threaded or ribbed steel glued-in rods (GiR), or by FRP plates [30], woven fabrics and rods (from glass, carbon, aramid, basalt, etc.). The elements used to substitute the decayed timber may be visible or not, and the elements that connect them to the sound part of the timber can be either external or internal [11].

The use of *prosthesis* became widely accepted mainly due to its low intrusion level, simplicity of design and execution, and the good aesthetic result of some of the used techniques. The type of *prosthesis* reinforcement method may vary, depending on many parameters (cultural values, aesthetics, presence of decorative elements, access to the damaged timber part, fire protection, on-site application, available expertise, cost, etc.).

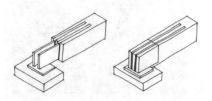

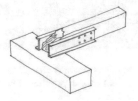

a) New timber elements connected to the sound part of the original timber with steel straps, plates and steel fasteners

b) New steel elements used to replace the decayed timber

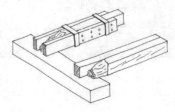

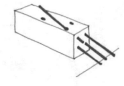

c) Glued-in plates connecting the prosthesis to the sound part of the original timber

d) Glued-in rods connecting the prosthesis made either of wood or resin

Fig. 16 Examples of end-beam repair techniques (*prosthesis*) using different materials and methods [11]

The above methods of *prosthesis*, especially the ones that use timber, offer several advantages: high connection stiffness without significant settling; possibility of ductile design with yielding of the steel or of the other types of bars in spite of the adherence based on glue; protection of the glue and the embedded elements from chemicals and fire; unmodified exterior of the reinforced element that maintains the original architectural characteristics [31]. A major advantage of the glued-in rod connections is the transfer of forces directly into the inner part of the members' cross-section [32].

For the use of external steel elements (sections or plates) and the problems that may arise, see Sect. 3.1.3 (Fig. 17). The use of timber *prosthesis* compared to resin or steel is considered as following more closely the conservation principles of historic timber structures [1].

The design of reinforcement for members or joints should take into account the effect that the reinforcement can have on the original structural system, which needs to be preserved, with the exception of course the cases that the original system presents important errors in the original conception (Figs. 17 and 18). Changes in beam or joint stiffness may have consequences for the overall behaviour and load distribution of the entire structure, altering the paths of loads and leading to their transfer to other parts of the structure.

Fig. 17 Interventions that may change the stiffness properties of the original semi-rigid joints [11]. In cases where steel is used, problems may arise due to environmental thermal or humidity changes on steel and timber

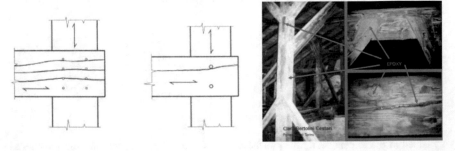

Fig. 18 Cracks caused by differential shrinkage parallel and perpendicular to the grain in the timber-to-timber connection (left) [33]. Improper filling of the shrinkage cracks with resin (right) [18]

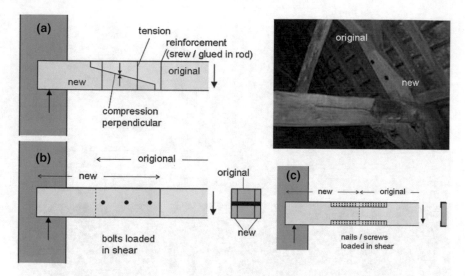

Fig. 19 Examples of end-beam repair techniques outside Hørsholm Church, Copenhagen, Denmark [34]

As the reinforcing elements generally have different stiffness and thermal expansion and moisture absorption properties than the timber elements, factors that constrain shrinkage and swelling due to thermal or moisture changes must be considered too, and if necessary, additional thermal or moisture-induced stresses should be accounted for in the design (Figs. 17 and 18) [10].

Some interventions have caused further damage to structures in the past, due to lack of knowledge on how to select and implement appropriate reinforcement methods. An example is the filling of shrinkage cracks with resins. The excessive stiffness of the adhesives and their subsequent inability to withstand the timber strains, especially strains due to hygrometric variations, can seriously impact the state of the existing cracking and even provoke new cracks (Fig. 18). The same applies to the use of bars glued to timber to stop the further widening of cracks. The use of such a technique can create undesired stress states, by preventing the natural movement of the timber.

Prosthesis Using New Timber Elements

As mentioned before, solid wood, glued-laminated timber, and laminated veneer lumber (LVL) can be used (Figs. 19, 20, 21, 22, 23, 24 and 25) as prostheses. Any new timber should have the same moisture content as the original one and, especially in cases where solid timber is used, the new timber should be of the same species as the old. An advantage of glued-laminated timber and LVL is that they have large sections. This feature enhances their use in historic structures, which in several cases

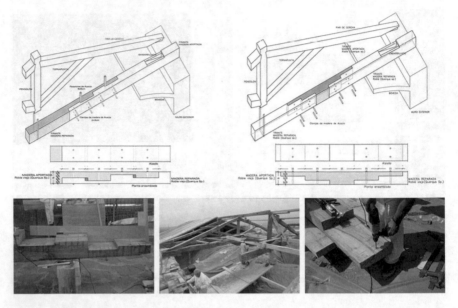

Fig. 20 *Prosthes*is using solid timber [35]

Fig. 21 Rinoji Temple in Japan: Different kinds of rafter repairs made by splicing damaged timber with new wood (left) [36]. Gekko-den Pavillion in Japan: Timber *prosthesi*s using the same timber species and the same techniques as the old building. The use of adze gives a smooth finish, matching the other timbers and ensuring similar aesthetic appeal and aging due to its superior resistance to moisture [37]

were built with large pieces of timber that nowadays are very difficult to be found [23].

In Japan, priority is given to repairs using traditional tools and techniques, hence carpenters repair timber structures by replacing rotten or damaged timbers with new timbers of the same species and dress them in the same way as the original (Fig. 21). In many cases, traditional handmade wrought iron nails are used in conservation work, especially in visible parts of the building, since their behaviour and appearance are different from those of modern fasteners. New elements should be marked with the date of the repair by making carvings on the wood. Several countries of Central and Northern Europe follow the same concept for the restoration of timber structures [7, 36].

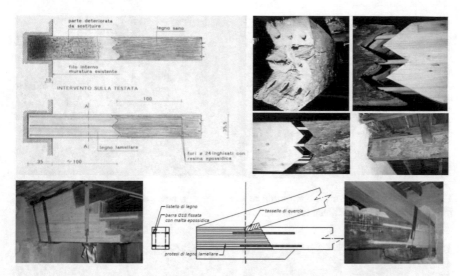

Fig. 22 Intervention on the historic roof of Valentino Castle (seventeenth century), Torino, Italy [27]. *Prosthesis* (glued laminated timber) at the end of the tie-beams and the rafters connected to the sound wood by fibre-glass rods

Fig. 23 Intervention on the historic roof of Clyne Castle in Wales (eighteenth century). The decayed timber was replaced in situ with a *prosthesis* made from laminated veneer lumber (LVL). Due to the restriction on access, the connection between the prosthesis and the hip rafters was set up using slots rooted in the sides of the *prosthesis* and the rafter, into which high tensile steel rods were bonded using a two-part epoxy structural adhesive [10]

Threaded rods enhance the perpendicular-to-grain strength of timber and provide an improvement in the plastic reserve of the beams and their adherence to the epoxy mortar. Glass fibre-reinforced polymer bars, in spite of their higher tensile strength, have a lower modulus of elasticity, resulting in larger deformations than those observed for steel reinforcements [39]. Due to their high strength properties, carbon fibre bars can provide lower diameters, compared to other types of rod, and offer a good solution in cases where the distances from the edges of the timbers are small (Fig. 24) [24, 25].

Fig. 24 Glued-in rods for the connection of old timbers to new ones. Threaded steel rods (left) [34]; carbon-fibre rods of small diameter [25]

Fig. 25 Timber roofs in the Arsenal of Venice, Italy. *Prosthesis* interventions using timber planks, screws and resin (courtesy C. Menichelli) [38]

Another method in which solid timber is used, is the replacement of a damaged section with timber boards (thickness 2 or 4 cm), of the same species as the sound wood, connected to each other and the sound timber by a mechanical system (self-threading stainless steel screws) and a bi-component resin (Fig. 25).

Prosthesis Using Epoxy Resin (EPX)

Epoxies were initially patented in the 1930s. However, it was not until the 1960s and 1970s that reinforcement techniques using epoxy resin were implemented in historic buildings [40]. The decayed part of a beam is replaced by a resin *prosthesis*, connected to the sound remaining part of the beam element, using glued-in plates (Fig. 26) or rods of different types, horizontal or inclined, in order to reinstate the mechanical properties of the timber element. The rods can be made of steel or FRP (glass, carbon or aramid); steel rods (galvanized or stainless) are preferred due to their reduced cost.

Fig. 26 Examples of prosthesis with epoxy resin [18]. Intervention using resins at the timber roof of Mons tower in Belgium

When the aesthetic value is of low practical relevance and no fire resistance is required, the resin part can be visible, or the connection of the prosthesis can be external (Fig. 26). The role of epoxy is two-fold: (i) it replaces decayed wood by functioning as a structural gap-filling adhesive (epoxy grout); and (ii) it adheres wood to the rods or plates (thixotropic epoxy resin) used for the connection of the resin part to the sound wood (Fig. 22, 23, 24). Viscosity is an important property to consider as consolidation repairs require very low viscosity to penetrate the wood cells, in contrast to gap-filling adhesives that require high viscosity to maintain their shapes and forms. In addition to viscosity, the working and curing times of an epoxy must also be taken into account. The most common resins used for timber reinforcement are polyurethane (PUR), pheno-resorcinol (PRF) and pheno-resorcinol-formaldeyde resins [41]. Longer working and curing times result in greater penetration. During application, the moisture content of the timber and the service temperature are critical factors that determinate the bond performance and durability. Appropriate bonding pressure of the timber components is not easily achievable on site, in contrast to factory situations where bonding is almost always carried out under controlled pressure. Adhesive manufacturers often specify the minimum and maximum application temperatures and pressures to be applied to the assembled components. For further information on this topic, see [41]. Models have been developed for strength prediction in tension and bending of *prosthesis* using epoxy resin. The bending capacity is, in relation to the strength of a full cross-section of sound wood, reduced considerably. Consequently, no intervention is applied to recover the full bending capacity. Since the polymer technique is mostly used at beam-ends, this is no problem (shear is most important at this location) [34].

Advantages of this method include its ease of use both locally and in situ, its low structural and fabric disruption, and its quick implementation since the materials are readily available. Another advantage is that the wood makes no contact with moist materials, e.g. masonry, thus reducing the probability of decay in the future. Special

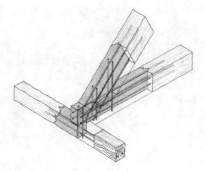

Fig. 27 Example of a rigid rafter-tie beam joint repair [15]. The occurring member is considered one single piece

attention must be paid to resin *prosthesis* applied to two or more timber members that form a connection, as in the heel joint of a roof truss. The joint becomes more rigid, causing important alterations in the overall structural system and leading to the transfer of loads to the weakest elements and consequently to damage or failures (Fig. 27).

Questions related to the compatibility of reinforcing and reinforced materials, the significant differences between the properties of wood, a hygroscopic organic material, and those of epoxy, an impermeable plastic, the durability, the low reversibility and behaviour of the resin under thermal and moisture fluctuations, and the long-term performance and fire resistance of the resin, are yet to be investigated. Epoxies are not recommended by timber experts for external repairs mainly due to moisture issues [7]. Moreover, some issues have prevented the wider use of adhesives, particularly in historical timber structures, where sufficient reliability cannot yet be guaranteed. One reason is that a long service life has not yet been fully proven for synthetic adhesives, since the oldest bonded joints are only around sixty years, and greater ages cannot be simulated by existing accelerated ageing tests.

Use of a New Structural Support System

In this case, a circumventing of the load path is achieved at the support area of the roof by the addition of a new element (steel, concrete, or stone). A disadvantage of the system is that an out-of-plane moment is transferred to the walls due to the eccentricity of the vertical loads transferred from the roof, since such new elements protrude from the walls (Fig. 28).

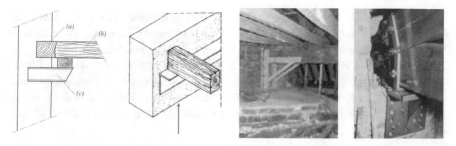

Fig. 28 Substitution of the support of decayed timbers with steel elements incorporated in walls, without necessarily removing the rotten parts. Intervention on a severely decayed rafter-tie beam connection (right), transferring the support to where the wood is sound using a properly designed steel element and applying an articulated collar to prevent sliding of the rafter [15, 19, 21]

3.2 Reinforcement for the Overall Load-Bearing System of the Roof

3.2.1 Addition of New Structural System in Parallel to the Original One

Rafters or tie beams, or even a whole roof truss, can be introduced between the main beams or the main trusses (placed parallel to them). An example of an added parallel system can be found in the main railway station in Wroclaw, Poland (Fig. 29, right) and in the roof of the Salone dei Cinquecento in Palazzo Vecchio in Florence (Fig. 29, left).

Roofing structures in several cases carry ceilings of great artistic value and the reinforcement may concern the increasing of the load-bearing capacity and mainly, the limiting of the deflection of the roofs, since in many cases the ceilings are supporting

Fig. 29 Salone dei Cinquecento in Palazzo Vecchio, Florence. General view of the two trusses for the roof at the back of the photo (original one) and the new one at the front (left) [42]. Main Railway Station in Wroclaw: a new load-bearing structure for the ceiling (a steel space frame) was used to relieve the roof truss from the coffer ceiling loads (right) [43]

frescos which are brittle without any ductility. This is the case of the king post timber roof designed in 1563 by Giorgio Vasari for the Salone dei Cinquecento in Palazzo Vecchio in Florence (project carried out 1563–1565, by Battista Botticelli), which had to carry a very heavy ceiling of great artistic value. The ceiling was soon affected by sagging caused by creep effects. A reinforcement work was carried out by Arch. D. Giraldi in 1854, who constructed new timber trusses placed, at a lower level, as a parallel system between the original ones without any intervention at the old ones (Fig. 29, left) [42].

3.2.2 Addition of New Structural Systems to Improve the Overall Performance and Stability of the Roof

Besides the behaviour of the individual structural elements (members and joints), which has an indirect but considerable impact on the overall behaviour of the system, reinforcement maybe needed to improve the original structural system that presents either design or execution errors, to confront a severe deformation of the whole roof, or an overall stability problem that needs bracing (Fig. 30). The advantage of these solutions is that they are reversible.

Another example of this type of reinforcement was used in Angera Castle (Lago Maggiore–Italy), where all loads are transferred after the intervention to the supports by a new steel truss, although originally the loads were divided across four supports [34].

Reinforcement of roofs to improve the overall load-bearing performance of buildings in seismic areas (use of different types of timber diaphragms or steel bracings and anchorages in masonry walls) has been discussed in Chapter "Seismic Reinforcement of Traditional Timber Structures".

 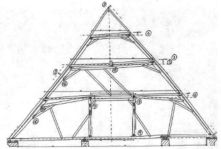

Fig. 30 St. Marien Church. Reinforcement of the timber roof using unbounded pin-loaded CFRP straps [44]

4 Conclusions

When new usage and/or new imposed loads are introduced, reinforcement of traditional timber structures is needed. If the decay of timber elements is great, then local replacement of the decayed part is the best solution. When interventions are necessary, specific and reliable on-site assessment techniques are required to determine the appropriate level of intervention needed. An important point remains the evaluation of the replacement, repair or reinforcing solutions, along with the cultural significance, the know-how, and the associated project costs of each case. Evaluation of the durability of the intervention work carried out with new innovative techniques is necessary too. The use of wood to solve the problems of wood offers one of the most interesting features in conservation, the compatibility.

For timber members, reinforcement helps to restore the load-bearing capacity that has been lost because of the material decay and to increase the resistance (strength and stiffness properties).

Current knowledge of reinforcement methods of existing structures is largely based on practical experience. Unfortunately, the study of reinforcement techniques is not yet included in European standards such as the Eurocode 5 (EN 1995-1-1) [28, 45], but only for specific aspects in the National Annexes of some countries.

Investigations on that promising topic have helped experts to figure out how to overcome timber weaknesses, resulting in proposals of design models and reinforcement methods. Some of the most important and applicable outcomes will be integrated into the revised edition of the Eurocode 5, to help engineers design reinforcements for timber structures. It must be pointed out though, that these reinforcements are for new timber structures.[2] Since existing structures and, mainly, historic structures are not covered by the new standards yet, it is urgent relevant European Standards to be developed too. Thankfully, in 2017, to close a part of the gap between practical needs and missing standardization, the European Standardization Committee responsible for the Eurocode 8, CEN/TC 250/SC 8 concerning earthquake design, decided to add in part 3 for existing buildings, a section on existing timber structures, historical or not.

Increased knowledge and research on retrofitting techniques is of great importance in order to support the Standards that are being or will be developed. When timber structures are reliably repaired or reinforced, structural failures and unnecessary replacements can be avoided, and sustainability, which is important from economic, environmental, historical and social perspectives, will be served too.

Standards, research, and the constant and continuous dissemination of the knowledge, mainly through education, can provide the necessary tools for structural engineers, who are often part of a multi-disciplinary team that have to work together in a restoration project, to evaluate the existing condition of a historic timber structure. Moreover, these tools may help in the selection of the proper interventions using

---

[2]For further information, see the relevant chapter on the RILEM state-of-the-art report and [1].

innovative and/or simple techniques that will sustain the authenticity of our architectural heritage, including the authenticity of the "invisible", in many cases, timber load-bearing structures.

References

1. IIWC ICOMOS Committee (1999) Principles for the preservation of Historic Timber Structures
2. ISCARSAH ICOMOS Committee (2003) Principles for the analysis, conservation and structural restoration of architectural heritage
3. Venice Charter (1964) Second International congress of architects and technicians of historical monuments, Venice, May 25–31
4. Yeomans DT (2008) Repairs to historic timber structures: changing attitudes and knowledge. In: DÁyala D, Fodde E (eds) Proceedings of the sixth international conference on structural analysis of historic construction. Preserving safety and significance, vol 1
5. Pinto L (2008) Inventory of repair and strengthening methods—timber. Advanced masters in structural analysis of monuments and historical constructions, Master thesis. University of Catalonia
6. COST FP1101 (2014) Site internet: www.costfp1101.eu
7. Larsen KE, Marstein N (2000) Conservation of historic timber buildings: an ecological approach. Butterworth-Heinemann, Oxford and Woburn (Mass.)
8. Cruz H, Yeomans D, Tsakanika E, Macchioni N, Jorissen A, Touza M, Mannucci M, Lourenço PB (2015) Guidelines for the on-site assessment of historic timber structures. Int J Arch Heritage Conserv Anal Restor 9(3)
9. UNI 11138 (2004) Cultural heritage—wooden artefacts—building load bearing structures—Criteria for the preliminary evaluation, the design and the execution of works. Ente Nazionale Italiano di Unificazione (in Italian), Milano
10. Franke S, Franke B, Harte A (2015) Failure modes and reinforcement techniques for timber beams State-of-the art. Constr Build Mater. https://doi.org/10.1016/j.conbuildmat.2015.06.021
11. Piazza M, Senno M (2001) Proposals and criteria for the preliminary evaluation, design and ecxecution of works on ancient load bearing timber structures. Wooden/handwork/wooden carpentry: European restoration sites. In: Bertolini-Cestari (ed) Proceeding of culture 2000 project: Italian action
12. Gasparini DA, da Porto F (2003) Prestressing of 19th century wood and iron truss bridges in the U.S. In: Huerta S (ed) Proceedings of the first international congress on construction history, Madrid, pp 978–986
13. Branco JM, Descamps T, Tsakanika E (2018) Repair and strengthening of traditional timber roof and floor structures. In: Costa A, Arêde A, Varum H (eds) Strengthening and retrofitting of existing structures. Building pathology and rehabilitation, vol 9. Springer, Singapore, pp 113–138
14. Tampone G (1996) Il restauro delle strutture di legno. Hoepli, Milano
15. Tampone G, Rugieri N, State-of-the-art technology on conservation of ancient roofs with timber structure. J Cultural Heritage 22
16. Gresford P (2009) Resin repairs to structural timber in building conservation in the UK. Thesis prepared for the architectural association conservation diploma course
17. Tsakanika E (2005) Methodology concerning the restoration of historical buildings. Case studies: the Turkish Mansion and the Hagi Mehmet Aga Mosque in Rhodes. In: Proceedings of international conference of the IIWC "conservation of historic wooden structures", Florence, vol 2, pp 194–203
18. Bertolini Cestari C, Marzi T (2018) Conservation of historic timber roof structures of Italian architectural heritage: diagnosis, assessment, and intervention. Int J Archit Heritage 12(4):632–665

19. Zamperini E, Corso di Recupero e Conservazione degli edifice. Criteri progettuali per il Consolidamento-Le strutture in legno. https://docplayer.it/6361795-Criteri-progettuali-per-il-consolidamento-le-strutture-in-legno.html
20. Tsakanika E, Kalafata P (2019) The timber roof of the Neoclassical School in the Medieval city of Rhodes island in Greece. The restoration project. In: Proceedings of 5th International Conference on Structural Health Assessment of Timber Structures, SHATIS'19, Guimarães, (Portugal)
21. Uzielli L (1995) Restoring timber structures—repair and strengthening. STEP 2 Timber Engineering, lecture D4. Centrum Hout, The Netherlands
22. Branco JM, Sousa HS, Tsakanika E (2017) Non-destructive assessment, full-scale load-carrying tests and local interventions on two historic timber collar roof trusses. Eng Struct 140:209–224
23. Lalane M, Pereira N, Valle A, Moraes P (2011) Projeto de intervenção em estrutura de madeira da cobertura da Igreja Matriz de São José, SC, Brasil. CIMAD 11-1º Congresso Ibero-LatinoAmericano da Madeira na Construção, 7-9/06/2011, Coimbra, Portugal
24. Hoath J. Repairing historic roof timbers. http://www.buildingconservation.com/articles/roofti mber/rooftimber.htm
25. Tsakanika – Theohari E, Mouzakis H (2010) A post-Byzantine mansion in Athens. The restoration project of the timber structural elements. In: Proceedings of 11th world conference on timber engineering, (WCTE 2010). Riva del Garda, Trentino, Italy, vol 2, pp 1380–1389
26. Branco JM, Tomasi R (2013) Analysis and strengthening of timber floors and roofs. In: Costa A, Miranda Guedes J, Varum H (eds) Structural rehabilitation of old buildings. Springer, pp 235-258. ISBN: 978-3-642-39685-4. URL: http://dx.doi.org/10.1007/978-3-642-39686-1, http://hdl.handle.net/1822/26659
27. Bertolini Cestari C, Invenizzi S, Marzi T, Spano A (2015) Numerical survey, analysis and assessment of past interventions on historical timber structures: the roof of Valentino Castle. In: Proceedings of international conference on structural health assessment of timber structures, SHATIS'15, Wroclaw (Poland), vol 2
28. EN 1995-1-1:2004, Eurocode 5: design of timber structures—Part 1-1: general   common rules and rules for buildings, +AC (2006) + A1 (2008) + A2 (2014), CEN European committee for standardization, Brussels, Belgium
29. Tsakanika E (2007) Byzantine and post-byzantine historical timber roofs in Greece. Typical failures, misunderstanding of their structural behaviour, restoration proposals. In: Proceedings of XVI international conference and symposium of the IIWC "from material to structure", Florence (http://www.icomos.org/iiwc/2007.htm)
30. CNR-DT 201/2005 (2007) Guidelines for the design and construction of externally bonded FRP systems for strengthening existing structures. Italian National Research Council, Rome
31. Piazza M, Tomasi R, Modena R (2005) Strutture in Legno. Milano, Italy, Ulrico Hoepli
32. Steiger R, Serrano E, Stepinac M et al (2015) Strengthening of timber structures with glued-in rods. Constr Build Mater 97:90–105
33. Piazza M, Tomasi R, Modena R (2005) Strutture in legno Hoepli Editore, Milano
34. Jorissen A (2012) Structural interventions. J Cultural Heritage 13S
35. Landa M (2005) REPARACIÓN DE ESTRUCTURAS DE MADERA MEDIANTE INJERTOS RECUPERACIÓN DE ELEMENTOS TRACCIONADOS. Artículo para la RE
36. Martinez A (2016) The current principles for the preservation of historic wooden monuments in Japan. In: Proceedings of the 20th international symposium of the IIWC, Falun, Sweden
37. Park TL (2015) Gekko-den case study: the process surrounding the preservation of historical wooden architecture in Japan. In: Toniolo L, Boriani M, Guidi G (eds) Built heritage: monitoring conservation management. Research for development. Springer, Cham
38. Menichelli C, Adami A, Balletti C, Bertolini Cestari C, Bettiol G, Biglione G et al. (2009) Le strutture lignee dell'arsenale di Venezia. Sudi et restauri. In: Proceedings of XXV Convegno Internazionale Scienza et Beni Culturali – Conservare e restaure il legno, Bressanone, Italy (in Italian)
39. Casals MV, Trujillo VR, Badia CL (2013) Timber structure repair of an emblematic catalan industrial building with wood grafts and epoxy resins. Adv Mater Res 778(2013):998–1005

40. Cleary R (2014) Considering the use of epoxies in the repair of historic structural timber University of Pennsylvania Philadelphia, PA. http://repository.upenn.edu/hp_theses/568
41. Pizzo B, Smedley D (2015) Adhesives for on-site bonding: characteristics, testing applications. In: State-of-the-art report on the Reinforcement of Timber Structures. COST Action FP1101
42. Pier Paolo Derinaldis, Gennaro Tampone. Rigidity versus ductility as an exception in timber structures planning in a moderately seismic area (the rigid timber trusses designed to strengthen the Vasari's Timber Ceiling in Palazzo Vecchio, Florence)
43. Bajno D et al (2013) Problems relating to assessment, repair and restoration of wooden roof structures in historic buildings, as exemplified by two case studies in Southern Poland. Adv Mater Res 778:888–894
44. Huster U, Brönnimann R, Winistörfer A (2008) Strengthening of a historical roof structure with CFRP straps. In: Motavalli M (ed) Proceedings of the 4th international conference on FRP composites in civil engineering (CICE2008), paper 8.B.4
45. Dietsch P (2017) Reinforcement of timber structures—a new section for Eurocode 5. In: Proceedings of the COST FP1402 conference "Connections—best practice", Graz, Austria

Retrofitting of Traditional Timber Floors

Luis C. Neves and Ivan Giongo

Abstract Timber floors are a critical component of many historical and modern buildings. Due to incorrect design and construction, effects of deterioration, change in use, and/or functional requirements, these components frequently need to be retrofitted. This chapter reviews the need for retrofitting of existing timber floors and critically evaluates a range of methods to improve their strength and stiffness, both in-plane and out-of-plane. The review of the available literature shows that this is a very active area of research and, in spite of the enormous progress achieved in the last decades, suggests that new and better methods will be developed in the future.

1 Introduction

Timber floors are a widely used solution in a range of buildings all over the globe, from medieval structures to modern single-family dwellings. From a structural viewpoint, these assemblies must support their own weight, other permanent loads, and the live load associated with their use. Moreover, in seismic regions, these floors serve as diaphragms, engaging the entire structure to resist horizontal forces. From a non-structural viewpoint, floors are critical for acoustic and thermal isolation, as well as for fire compartmentalization. The most common floor typology is the so-called "straight sheathing", where a single layer of timber boards (forming the decking) is fixed perpendicularly to the timber joists (forming the framing) via iron/steel nails or wood pegs.

In older structures, existing floors were built based on empirical knowledge, without any formal engineering design. This, combined with deterioration, increased live load, and, above all, more stringent requirements in terms of safety and comfort

L. C. Neves (✉)
Department of Civil Engineering, University of Nottingham, Nottingham, UK
e-mail: Luis.Neves@nottingham.ac.uk

I. Giongo
Department of Civil, Environmental and Mechanical Engineering, University of Trento, Trento, Italy

© RILEM 2021
J. Branco et al. (eds.), *Reinforcement of Timber Elements in Existing Structures*,
RILEM State-of-the-Art Reports 33,
https://doi.org/10.1007/978-3-030-67794-7_11

of users, has led to the need to retrofit a large number of existing timber floors. The key properties to consider in the design and retrofit of timber floors are: (i) out-of-plane bending strength, to guarantee the safety of occupants under vertical loads; (ii) out-of-plane bending stiffness, to limit deflections and vibrations under vertical loads and, thus, increase the comfort and sense of safety of users; and (iii) in-plane strength and stiffness, to guarantee adequate transfer of loads across all structural elements under horizontal loads (e.g. wind and earthquake). In this chapter, retrofitting is understood as a range of techniques which, more than restore the performance of the structure to its original condition, aim to achieve a higher level of performance than the initial construction.

A range of techniques have been developed over the last decades to improve in-plane and out-of-plane stiffness and strength of timber floors. The variety of structures, their use and their historical value require the use of very different approaches. If an overlay of concrete is a low-cost alternative and dramatically increases both in-plane and out-of-plane strength and stiffness, it has serious aesthetic impact which makes it incompatible with buildings with historical value. At their opposite ends, floors can be retrofitted using a fully reversible timber-based approach, which, however, is more expensive, requires more experienced workmanship and results in smaller increases in strength and stiffness. The choice of the ideal retrofit technique requires a clear understanding of the existing possibilities, but also a careful evaluation of the restrictions present in each case [1–3]. These include the historical value of the building and the need to conform to the Venice Charter in terms of reversibility of actions, the available working space, the available free height, accessibility of the upper and lower face of the pavement, amongst others.

In the following sections, the fundamental retrofitting techniques for existing timber floors will be described, their advantages and disadvantages discussed, and models and design approaches presented. First, retrofitting techniques applicable to common structures will be evaluated. These tend to focus on low costs, while sacrificing reversibility and, sometimes, aesthetics. After this, methods which aim at partial or full reversibility will be discussed, including the use of traditional methods and materials.

2   Performance and Requirements of Existing Timber Floors

As described above, retrofitting of timber floors aims at improving the performance of these elements in out-of-plane bending and shear and, in seismic regions, in in-plane behaviour.

2.1   Out-of-Plane Performance

The out-of-plane bending retrofitting is relevant for both ultimate limit states and service limit states. In fact, many existing floors have, in the course of their lives, suffered the effects of age such as timber decay and biological attack, thus reducing the cross-sections of their structural elements. Depending on the severity of deterioration, this might require the reinforcement of the element, its complete replacement, or replacement of some parts (Fig. 1).

Besides the effect of deterioration, it must be considered that users of buildings are significantly more demanding nowadays in terms of comfort, aesthetics and safety. For this reason, excessive deformation, due to both permanence and live loads, and vibrations are much less acceptable today than when many of existing buildings were initially built. The natural frequency of timber floors can be computed as (EC5):

$$f_1 = \frac{\pi}{2l^2} \sqrt{\frac{(EI)_l}{m}} \tag{1}$$

where m is the mass per unit area, in kg/m^2, l is the floor span, in m; and $(EI)_l$ is the equivalent plate bending stiffness of the floor about an axis perpendicular to the beam direction, in Nm^2/m. Floors with frequencies below 8 Hz are likely to experience resonance under normal use loads, and should be avoided. This can only be achieved by increasing the stiffness of the pavement, as reducing the mass is extremely difficult, especially for timber floors.

Increasing the strength and stiffness of floors in out-of-plane bending is usually achieved by the addition of new materials, thus creating a new composite section. Timber, concrete, steel and FRP are used in different cases, with different structural, aesthetic and financial consequences. In the design of such composite sections, the bond between the different materials plays a crucial role, and extensive research has been dedicated to the design and characterization of connections using timber and steel connectors, as well as a range of adhesives.

(a)   (b)

Fig. 1   Images of a severely deteriorated historical timber diaphragm (courtesy of Maurizio Piazza)

2.2 In-plane Performance

Many existing single-family dwellings built before the development of modern seismic codes have non-reinforced masonry and timber floors/roofs, even in areas of high seismic risk. The performance of these structures under seismic loads is very deficient, as shown, for example, by the 1989 Loma Prieta Earthquake in California [4], the 1997 Umbria-Marche earthquake in Italy [5, 6] and, more recently, in New Zealand in 2011 [7] and again in Italy in 2009, 2012 and 2016 [8–10].

The analysis of the performance of these buildings during earthquakes has shown that the main cause of failure is the inability of the timber floors/roofs to work as diaphragms, connecting all walls and ensuring that seismic loads are carried by walls responding in-plane, rather than out-of-plane. There are two main reasons for this: (i) lack of in-plane stiffness and strength of the timber diaphragms and (ii) inadequate connections between the floors and walls. In fact, frequently, the timber floor is connected to the walls simply by having the joists sit on pockets in the walls. If this is reasonable to carry vertical loads, it provides no real connection under seismic loads. The in-plane deflection of the timber diaphragm can induce tension and compression zones (diaphragms with larger aspect ratios tend to exhibit a more pronounced in-plane flexural response, see Fig. 2). If compression stresses can be resisted by the timber decking, tension can be critical and, consequently, specially designed connections between the diaphragm and wall are often employed in refurbishment interventions. Baldessari et al. [11] analysed the use of steel chords and concrete ring-beams around the perimeter of timber floors and their combination with other reinforced techniques as a means to improve the diaphragm behaviour of floors (Fig. 3). In this regard, it is worth mentioning the "girder analogy" suggested by ATC in 1981 [12], where the chords of a diaphragm represent the flanges of the girder and resist bending forces and the sheathing represents the web of the girder and absorbs shear forces. A specific remark must be made on straight-sheathed diaphragms. In the case of single straight sheathing, timber diaphragms tend not to exhibit a clear

Fig. 2 Deformation of a timber floor diaphragm under horizontal loads [11]

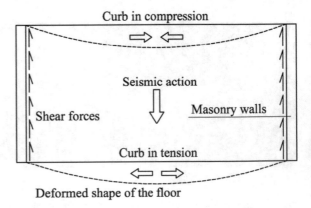

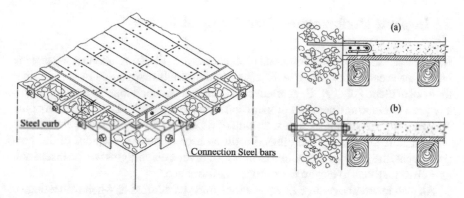

Fig. 3 Wall-to-diaphragm connections. Connection **a** between concrete ring-beam and the walls, and **b** between steel ring-beam and the walls [11]

flexural response, which limits the ability of the diaphragm to engage the contribution of the chords to the overall diaphragm stiffness (for more details refer to [13]). It is also important to note that diaphragm perimeter chords can confine the masonry and consequently contribute positively to the masonry spandrel rocking capacity and the overall seismic response of the building. The use of concrete ring-beams should, however, be evaluated carefully, as excessively stiff ring-beams can induce detrimental effects on the masonry walls due to concrete-masonry deformation incompatibility [14]. Furthermore, the use of concrete ring-beams at intermediate building levels, where the removal of masonry portions to create spaces to support the concrete beams is required, should be avoided [15].

The lack of diaphragm in-plane strength and stiffness results in out-of-plane deformation of the walls beyond their stability limit, leading to failures such as those shown in Fig. 4.

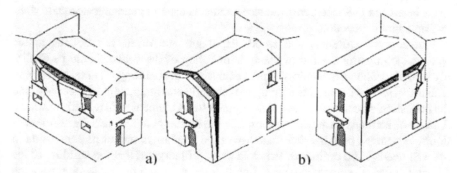

Fig. 4 Failure of masonry walls during earthquake: out-of-plane I° mode mechanisms associated with inadequate diaphragm strength and stiffness and wall-to-diaphragm connection [16]

3 In-plane Performance Retrofitting of Floors

As discussed above, timber floors and roofs can serve as diaphragms, ensuring that the inertia loads caused by earthquakes are carried by walls responding in plane, rather than out of plane [17–19]. To guarantee this, it is fundamental that the diaphragm has adequate in-plane stiffness and proper connection to the walls. The evaluation of the in-plane stiffness of timber floors, as well as its impact on the seismic performance of unreinforced masonry buildings, is still an area of active research. In the next paragraphs, the properties required of floors to serve as diaphragms will be discussed, as well as methods to ensure their proper performance.

As mentioned above, two key properties must be addressed when evaluating or designing floors in seismic areas: the in-plane stiffness and strength of the diaphragm and the connection between the diaphragm and the walls. As discussed in Brignola et al. [20], excessively stiff diaphragms with inadequate connections to the walls can result in undesired failure modes associated with both torsion modes and expulsion of the corners, as observed in the Umbria-Marche earthquake (1997). Also, Nakamura et al. [21] and Scotta et al. [22] have shown that in some cases, stiffer diaphragms may lead to higher seismic demands on the masonry walls compared to more flexible diaphragms.

Considering that in most existing buildings there are no connections between the diaphragms and the walls, this frequently needs retrofitting. A technique that was frequently used to enhance the connection between the two elements was the insertion of concrete ring-beams in the masonry wall. However, the weakening of the masonry wall that resulted from this has been observed to cause significant increase in the risk of failure under earthquakes, deeming this approach unsatisfactory [14, 23]. The existing literature proposes a range of techniques based on the use of steel elements to connect floors and walls [24, 25]. The main advantages of these techniques are their simplicity, low impact on the wall, continuous connection between the diaphragm and wall, and ductility. Figure 5 shows examples of steel ties parallel or perpendicular to the beam and L-Shape perimeter steel elements used to guarantee adequate wall-to-diaphragm connection.

The in-plane stiffness of a wood diaphragm and of its connections to the walls, which control the out-of-plane deformation of the walls, should be explicitly modelled when a dynamic or modal analysis of the building is being carried out. Unlike for other materials, the assumption of a rigid diaphragm is rarely accurate, and the floor should be modelled as a slab or plane stress element with realistic stiffness. A range of methods have been proposed in the literature to evaluate the stiffness of floors. In general, both the deformability of the diaphragm and of its connection to the wall needs to be evaluated. Brignola et al. [26] suggest that the stiffness of the diaphragm can be computed considering that the floor and the connection between the floor and the wall are a series system, so that:

$$\frac{1}{k_{eq,c+d}} = \frac{1}{k_{eq,d}} + \frac{1}{k_c} \qquad (2)$$

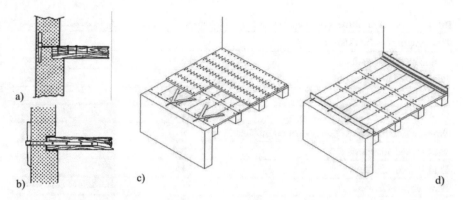

Fig. 5 Techniques of improving the connection between a timber floor and masonry walls using metallic elements: **a, b** steel ties; **c** steel ties perpendicular to beams way; **d** L-shape perimeter steel element [20]

where $k_{eq,c+d}$ is the equivalent stiffness of the floor system, $k_{eq,d}$ is the equivalent stiffness of the diaphragm, and k_c is the stiffness of the floor-wall shear connectors. Brignola et al. [26] presents analytical expressions for simple floors, considering the deformability of timber elements and of connections between elements. For example, for a single straight sheathing, the shear stiffness can be computed as:

$$G_{eq} = \frac{\chi}{A} \cdot \left(\frac{1}{k_{ser} s_n^2} + \frac{1}{GA} + \frac{l^2}{12EI} \right) \tag{3}$$

where k_{ser} is the nail deformability that can be determined experimentally through testing (see [27] and [28]) or using experimental formula provided by, for example, the Eurocode 5 [29]; χ is the shear factor; G is the shear modulus of timber planks; E is the flexural modulus parallel to the grain of timber planks; A is the area of plank section; I is the moment of inertia of plank section; l is the wheelbase between beams; and s_n is the spacing between nails.

A simplified approach is proposed in ASCE/SEI 41-17 [30], where values for inplane stiffness to be used in the linear elastic analysis of buildings are provided for a set of diaphragm types, considering both non-retrofitted and retrofitted cases. Table 1 shows some of the values proposed in this document with respect to the shear stiffness (G_d). The equivalent shear stiffness of each diaphragm type is defined as the shear modulus multiplied by the diaphragm thickness. The experimental background for the values reported in Table 1 goes back to the early 1980s and the ABK test program [31]. New Zealand standard NZSEE [32] provides equivalent stiffness values for straight-sheathed diaphragms based on the experimental outcomes from Wilson et al. [33] and Giongo et al. [34]. Such stiffness values, shown in Table 2, are secant values measured at a diaphragm deformation related to the onset of out-of-plane failure of face-loaded masonry walls. The NZSEE suggests different G_d values for loading direction parallel and perpendicular to the joist orientation and for different

Table 1 Expected stiffness (G_d) values for some diaphragm types according to ASCE/SEI 41-17

Diaphragm type G_d (kN/m)	
Single straight sheathing	350
Double straight sheathing	
Chorded	2670
Un-chorded	1240
Wood structural panels overlay on straight sheathing	
Un-blocked, un-chorded	870
Un-blocked, chorded	1580
Blocked, un-chorded	1240
Blocked, chorded	3200

Table 2 Shear stiffness values for straight-sheathed vintage flexible timber floor diaphragms, according to [32]

Direction of loading	Joist continuity	Condition rating	Shear stiffness $(G_d)^a$ (kN/m)
Parallel to joists	Continuous or discontinuous joists	Good	350
		Fair	285
		Poor	225
Perpendicular to joists[b]	Continuous joist or discontinuous joist with reliable mechanical anchorage	Good	265
		Fair	215
		Poor	170
	Discontinuous joist without reliable mechanical anchorage	Good	210
		Fair	170
		Poor	135

[a]Values may be amplified by 20% when the diaphragm has been re-nailed using modern nails and nail guns
[b]Values should be interpolated when there is mixed continuity of joists or to account for continuous sheathing at the joist splice

condition ratings, from "Poor" to "Good". For further details on the assessment procedures proposed by the ASCE and NZSEE and their similarities/differences, see Giongo et al. [35, 36].

In the last two decades, researchers have proposed and tested several solutions to improve the in-plane behaviour of timber diaphragms. Such solutions differ from each other in many aspects, including the in-plane increase in performance, the level of invasiveness, the reversibility, and the costs. Therefore, as already mentioned, there is no such thing as the "best technique" and the optimal choice needs to be evaluated case by case. The solutions range from additional sheathing made of wooden boards [11, 37–39] to nail plates that limit relative movement between floorboards [40, 41]; from steel straps [11, 39, 42, 43] and FRP straps [11, 39, 41, 43, 44] that enable strut-and-tie mechanisms to a more "slab-like" response ensured by plywood/OSB panel overlays [11, 33, 45–47] or by a CLT panel overlay [47, 48].

Figure 6 shows a summary by Schiro et al. [49] of the most relevant techniques, with a qualitative rating of their performance addressing the increase in in-plane and out-of-plane stiffness and the added mass associated with each strengthening intervention.

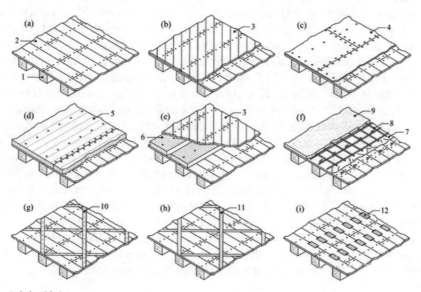

1. Original joist
2. Original floorboards
3. Additional diagonal sheathing
4. Structural wood-based panel
5. Cross laminated timber (CLT) panel or laminated veneer lumber (LVL) panel
6. Glulam or LVL plank
7. Waterproof sheath
8. Steel reinforcement
8. Steel reinforcement
9. Concrete slab
10. Metal straps
11. FRP – CFRP straps
12. Nail plates

Sol.	Stiffness increase		Lightweight
	Out-of-plane	In-plane	
b)	○○○○○	●●●○○	●●●●○
c)	○○○○○	●●●○○	●●●●○
d)	●●●●○	●●●●○	●●●○○
e)	●●●●○	●●●○○	●●●○○
f)	●●●●●	●●●●●	●○○○○
g)	○○○○○	●●○○○	●●●●●
h)	○○○○○	●●○○○	●●●●●
i)	○○○○○	●○○○○	●●●●●

Fig. 6 Diaphragm strengthening and stiffening solutions adapted from [49]: **a** original floor; **b** additional diagonal sheathing; **c** structural wood-based panels; **d** CLT/LVL panels; **e** timber planks and additional diagonal sheathing; **f** concrete slab; **g** metal straps; **h** FRP/CFRP straps; **i** nail plates

4 Out-of-Plane Performance Retrofitting of Floors

4.1 Timber-to-Concrete Composites (TCCs)

A technique commonly used for retrofitting existing timber floors is the in situ introduction of a cast concrete overlay (Fig. 7). This new layer is connected to the existing
timber using a variety of steel connectors and/or notches on the wood. When adequate
bonding between concrete and timber is achieved, a composite section is created,
wherein the concrete layer works in compression and the timber joists are subjected
to tension. This technique has relevant advantages, including: (i) significant increase
in strength and stiffness in out-of-plane bending relative to the timber-only floor, (ii)
increase in in-plane stiffness; (iii) considerable acoustic separation; (iv) increased
thermal mass and isolation; as well as (v) increased fire resistance. By making use
of the existing timber floor deck as formwork, the construction process is fast and
requires limited machinery. However, there are also significant drawbacks which limit
the applicability of this technique. On the one hand, when an overlay of concrete is
added to existing buildings, there is a significant increase in mass which not only
increases the vertical loading on walls and foundations but also the loading due to
seismic action [50]. Lightweight concrete has been touted as an alternative that minimizes this effect while keeping most advantages of the composite concrete-timber
solution [51]. On the other hand, this is a non-reversible retrofitting technique, as
demolishing the concrete slab without damaging the timber floor is almost impossible. This, combined with the aesthetic effects of using concrete in an existing timber

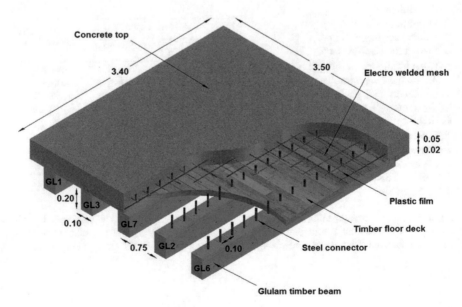

Fig. 7 Composite concrete-timber slab (adapted from [52])

structure, has significantly reduced the usefulness of this approach on buildings with historical value. Other, albeit minor, limitations include the increase in the thickness of the slab and the consequent reduction in free height.

Several design methods have been proposed in the literature, with a simplified approach proposed in the Eurocode 5, even if only timber-concrete composite bridge decks have been considered. The design of these elements is complex due to two phenomena: (i) the partial composite action that arises from the flexibility of the timber-concrete connection and (ii) the different time-dependent behaviour of the materials (concrete, timber, and steel connectors) involved. Two sets of approaches have been explored in the literature to design and predict the performance of these composite systems. The first is the basis of the method proposed in the Eurocode 5. In this case, the behaviour of all components of the system are assumed linear up to failure. This is a reasonable assumption if the expected failure mode is crushing of the concrete slab or failure of the timber joists without yielding of the connectors, which is often the case when strong connectors are used. This method, often referred to as the gamma method (originally proposed by Mohler [53]), can be applied to a range of composite connections, as will be discussed later. The fundamental principle of this method is the definition of a bending stiffness $(EI)_{ef}$, which is the sum of the stiffness of the timber and concrete elements in isolation, $E_c I_c$ and $E_t I_t$, and an increase in stiffness due to the composite behaviour $(\gamma_c E_c A_c a_c^2$ and $\gamma_t E_t A_t a_t^2)$ as:

$$(EI)_{ef} = E_c I_c + E_t I_t + \gamma_c E_c A_c a_c^2 + \gamma_t E_t A_t a_t^2 \qquad (4)$$

where E, I, A and a respectively represent the modulus of Young, moment of inertia, area and distance between centroid and neutral axis, for concrete (subscript c) and timber (subscript t). γ represents the shear connection reduction factor, which depends on the connection between elements. Following the Annex B of the Eurocode 5, the gamma factors can be estimated as:

$$\gamma_C = \frac{1}{1 + \frac{\pi^2 \cdot E_C A_C s_{ef}}{K l^2}} \qquad (5)$$

where s_{ef} is the effective spacing of the connectors assumed as smeared along the span of the floor beam; l is the span of the timber-concrete composite (TCC) floor beam; and K is the slip modulus of the connector. The slip modulus is a linear approximation to a non-linear force-slip relation. To improve the accuracy of this estimation, it is frequent to consider an in-service (K_s) and ultimate (K_u) moduli, each fitted to a different region of the force-slip curve. These values can be evaluated using push-out tests, for example by following EN 26891 thus:

$$K_s = \frac{0.4 F_m}{v_{0.4}} \qquad (6)$$

$$K_u = \frac{0.6 F_m}{v_{0.6}} \qquad (7)$$

where $v_{0.4}$ and $v_{0.6}$ are the slips at the concrete-timber interface under a load of 40% and 60% respectively of the mean shear strength F_{m0}.

The elastoplastic approach has been used when failure of the timber beam occurs after plasticization of the connectors. In this case, the connection is assumed to have a perfectly elastic plastic behaviour, and strain and stress can be computed assuming the bond is fully plastic. Although more accurate [54], this method is also more complex. The stiffness of the connection between concrete and timber depends strongly on the method used to connect the two components. As shown by Yeoh et al. [55], dowel-type connectors introduce a low-stiffness and low-strength connection, which increases when metal plates or notches are introduced in the timber element [56, 57] (see Figs. 8 and 9). The stiffness of dowel-type fasteners can be significantly improved if the axial stiffness of the threaded screws is exploited by arranging them in an X-type configuration, as shown by Van der Linden [58], Dias and Jorge [59] and Sebastian et al. [60, 61], just to name a few authors.

Regarding the impact of time-dependent phenomena, like shrinkage and creep of concrete, a first method was proposed by Ceccotti [62] based on the use of the effective moduli for timber, concrete and bond in the form:

$$E_{ef} = \frac{E}{1 + \varphi} \tag{8}$$

where φ is the pure creed coefficient of each material at the end of its service life. This approach does not account for the internal forces and deformations caused by differential environmental strains. Frangiacomo [63] proposed a close-form expression for taking these effects into account. A simpler approach was proposed by Schanzlin

Fig. 8 Comparison of different categories of connection systems (adapted from [55])

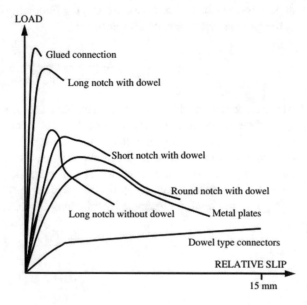

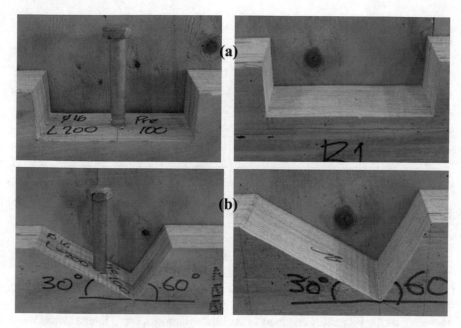

Fig. 9 Examples of indentations and combinations of indentations with connectors aimed at improving the strength and stiffness of the bond between timber and concrete (adapted from [55])

[64] using the expression above, replacing the pure creep coefficient by an empirical effective creep coefficient.

A detailed review of the several solutions and design techniques for TCCs proposed by researchers over the last few years is provided by Dias et al. [65, 66].

4.2 FRP Retrofitting

Fibre-reinforced polymers are composite materials made up of high-strength and-stiffness fibres that are embedded in a polymeric resin, which transfers loads across the fibres and protects them. The fibres are composed of materials with very high strength and stiffness but usually very brittle behaviour, including carbon, e-glass, basalt and aramid (Table 3). By using these materials in a fibre form, it is possible to reduce the impact of the defects on strength, increasing both the mean strength and the reliability (i.e. by reducing variability), which allows their use in structures. The polymeric matrix is a resin, usually polyester, vinyl-ester, polyurethane or epoxy. Its main role is to protect the fibres from external aggression and bond them together.

FRP is available both as a wet setup, where the matrix and fibres are applied separately and the matrix is cured on site, and pre-cured, usually as pultruded plates or rods. FRP has been widely used for the retrofit of reinforced concrete structures

Table 3 Fibre and polymer properties from Hollaway and Teng [70] and Greco et al. [71], as shown in Schober et al. [69]

Material	Modulus of elasticity (GPa)	Tensile strength (MPa)	Failure strain (%)	CTE (10^{-6} $°C^{-1}$)	Density (g/cm^3)
E-glass	70–80	2000–4800	3.5–4.5	5.0–5.4	2.5–2.6
Carbon (HM)	390–760	2400–3400	0.5–0.8	−1.45	1.85–1.90
Carbon (HS)	240–280	4100–5100	1.60–1.73	−0.6 to −0.9	1.75
Aramid	62–180	3600–3800	1.9–5.5	−2.0	1.44–1.47
Basalt	82–110	860–3450	5.5	3.15	1.52–2.7
Polymer	2.7–3.6	40–82	1.4–5.2	30–54	1.10–1.25

CTE coefficient of thermal expansion; *HM* high modulus; *HS* high strength

over the last decades. More recently, its potential for retrofitting timber elements, in particular beams, has led to intensive research work.

FRP is relatively small in size, which means that significant improvements in strength and stiffness can be achieved without a significant change in geometry. However, as the FRP must be glued to the timber element, these interventions must be regarded as non-reversible or partly reversible and, consequently, their use in structures with historical value must be carefully considered. A range of adhesives have been used to glue FRP to timber elements, including epoxies, polyurethanes, polyesters, phenolics and aminoplastics [67, 68]. However, of these, two-part cold-cure epoxy adhesives have been found to be more reliable on site [69].

FRP can be used in a range of scenarios to retrofit timber structures. In the next paragraphs, some of the most common use in the retrofit of timber floors will be discussed briefly. The main impact of FRP is on out-of-plane behaviour. FRP can be used to increase strength in bending, shear and to connect different elements or element parts.

The most frequent use of FRP is in the bending reinforcement of beams. Experimental studies [72–78] have shown that even small amounts of FRP reinforcement can increase the strength and stiffness of solid timber and glulam beams up to 90% and 100% respectively [69]. The reinforcement of beams in bending is usually achieved by bonding FRP plates to the bottom face of the beam (externally bonded reinforcement) or in grooves cut near the bottom of the beams (near-surface mounted reinforcement). As for other retrofitting methods, the strength and stiffness of the bond between the FRP and timber element is critical to guarantee a composite behaviour, with full engagement of the FRP and prevention of premature failure due to debonding. In recent years, extensive research has focused on the evaluation of the prediction of the FRP-timber bond performance. In order to design beams retrofitted with FRP, it is crucial to consider all possible failure modes. Considering the difference in ultimate strains for timber and FRP, failure of the FRP is unlikely, and the failure modes that need to be taken into account are [69] failure of timber in tension while timber in compression remains in the elastic region and failure in tension after the onset of compressive yielding.

For the first case, the ultimate bending moment can be computed, considering a linear distribution of stresses, as shown in Fig. 10a [69]. In this case, the ultimate bending moment is:

$$M_u = F_f \cdot \left(h_{NA} - h_f\right) + \frac{2}{3}F_{cw} \cdot (h - h_{NA}) + \frac{2}{3}F_{tw}h_{NA} \qquad (9)$$

where F_f is the force in the FRP, h_{NA} is the position of the neutral axis, F_{cw} and F_{tw} are the respective resultants of compression and tension stresses in the wood. The position of the neutral axis can be computed by solving the equation:

$$(E_w - E_f)A_f\frac{h_{NA} - h_f}{h - h_{NA}} + \frac{1}{2}E_wb(h - h_{NA}) - \frac{1}{2}E_wb\frac{h_{NA}^2}{h - h_{NA}} = 0 \qquad (10)$$

In the second case (Fig. 10b), the strain in compression is higher than the yield strain. In this case, the ultimate moment is given by:

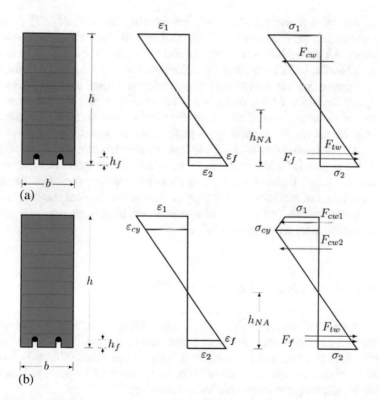

Fig. 10 Distribution of stresses in timber section reinforced with FRP

$$M_u = F_f \cdot \left(h_{NA} - h_f \right) + \frac{2}{3} F_{cw1} \cdot h_{cy1} + \frac{2}{3} F_{cw2} \cdot h_{cy2} + \frac{2}{3} F_{tw} \cdot h_{NA}$$

Ideally, the beam should be unloaded before application of the FRP reinforcement. If this is not the case, the existing strain in the structure before the application must be taken into account in the diagrams in Fig. 10.

One of the main limitations of the use of FRP as a retrofit material, particularly in Europe, is the absence of a European Standard for the design of these structures, especially in regard to the definition of design values and partial safety factors for these materials. It is expected that, in the near future, FRP will be included in the Eurocodes, building on existing experience in coding mostly in Italy [79].

When subjected to elevated temperatures, the strength and stiffness of FRP are severely reduced due to the loss of structural integrity of the polymer matrix [80]. Moreover, epoxy adhesives used for bonding FRP to timber have glass transition temperatures of up to 100 °C, meaning that externally bonded FRP will suffer both loss of reinforcement strength and loss of bonding when exposed to fire. Martin and Tingley [81] embedded FRP in a timber element as a means to reduce its exposure to elevated temperatures, showing that a 44% increase in fire endurance could be achieved.

To further enhance the effect of FRP, pre-stressing can be used [82]. In fact, in passive reinforcement, failure occurs in the timber while the FRP is under relatively low stresses. An alternative is to apply a tensile force to the FRP sheet or bond by means of hydraulic jacks before bonding. The eccentricity of the FRP relative to the neutral axis causes stresses and deformations that oppose those caused by the vertical loads. A key limitation of this technique is the need to anchor the pre-stress force to the end of the beam, which can cause delimitation due to the high concertation of stresses. Moreover, in retrofitting projects, space for the hydraulic jacks might not exist. An alternative method [83] consists in pre-cambering the timber element before bonding the FRP reinforcement. By introducing a triangular bending moment diagram in the beam, this method leads to a low and constant shear stress along the glue line. A range of methods for applying pre-stress in a gradual manner, reducing the stress concentration and the risk of delamination, have been proposed in recent years, as described by Kliger et al. [84] and.

4.3 Timber-to-Timber Composites (TTCs)

When retrofitting historical timber floors, it is critical to limit the use of modern materials and guarantee that all interventions are reversible. To achieve this, the main structural material must be timber or timber derivatives. Moreover, connections between existing and new elements or materials must be achieved using dry connections, allowing for disassembling if necessary.

A retrofitting technique that satisfies these constrains while significantly increasing both strength and stiffness is the addition of timber planks to an existing joist, connecting both using bolts, screws or wood dowels. If adequate connection is achieved between the old and new elements, they form a new composite beam (see Fig. 11), with a very large increase in the moment of inertia and, consequently, in the stiffness and strength. First examples of timber-to-timber strengthening solutions can be found in Piazza [85], where the reinforcing "wooden slab" (made of timber boards or sheets) was fixed to the existing joists through the original sheathing with epoxy-injected threaded steel rods (Fig. 12). Modena et al. [86] proposed an alternative solution where dry connection with wood dowels is employed and the reinforcing timber planks are separated from each other by a gap. The gap between planks allows the introduction of technical installations and thermal and acoustic isolation solutions without increasing further the floor thickness. On the other hand, if the in-plane stiffness of the diaphragm needs to be increased, an extra layer of diagonal boards (or plywood/OSB panel sheathing) is required. Valluzzi et al. [87] applied CFRP sheets at the bottom surface of existing joists to further increase the flexural performance of the solution proposed by Modena et al. [86] (Fig. 13).

To improve the distribution of stresses between the upper flange and the web, the existing floor should, when possible, be propped before application of the retrofit. This ensures that the connection between the old and new elements is executed while both are in the undeformed state. Dry connection systems have become more and more popular thanks to the development of many types of timber screw fasteners

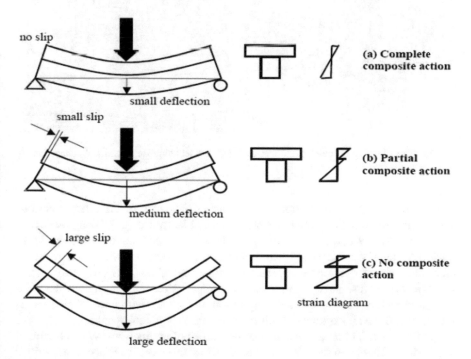

Fig. 11 Composite actions (adapted from [55])

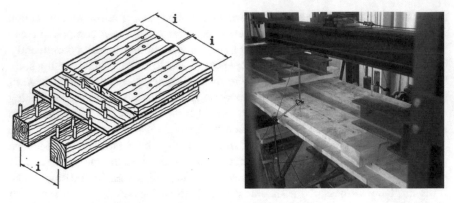

Fig. 12 First examples of timber-to-timber solutions [85]

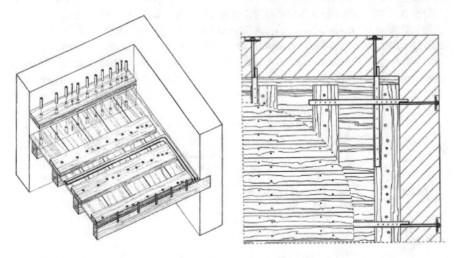

Fig. 13 Strengthening of floor with timber-to-timber dry technique and details of connections to masonry walls [86]

suitable for manifold applications. Experimental evidence of the effectiveness of screw connections for TTCs has been presented by Piazza et al. [88], Giongo et al. [89] and Riggio et al. [90], where the test results show TTC specimens (reinforcing planks made by glulam beams arranged on their side) with a flexural behaviour close to that of an ideal composite section (i.e. connection efficiency values > 0.7), especially when the fasteners are inserted in a shear-tension configuration, with an angle to the grain of approximately 45°. In such a configuration, in fact, the withdrawal resistance of the fastener thread is exploited, resulting in a remarkably higher connection capacity and stiffness (insight on the mechanical properties of inclined screws can be found in [91–93]). Thanks to their helical thread, screw fasteners for timber connections can generate a significant compression force (up to values > 10 kN; see Giongo et al. [94]) between the elements they connect. This compression force can be exploited to pre-stress and pre-camber existing timber joists [95–97], thereby improving the out-of-plane performance of TTCs. Such a pre-stressing/pre-cambering procedure depends on the correct sequential insertion of the fasteners but does not require propping systems or external tendons (see de Lima et al. [98] for a recent example of beams pre-stressed via external tendons) (Fig. 14).

Over the years, timber planks have gradually been replaced by CLT panels [96, 99], which allow for simultaneous increase in both the in-plane (see Sect. 3) and out-of-plane diaphragm performance, provided that an effective slab-to-slab connection is present [100, 101]. Roensmaens et al. [102, 103] have proposed the insertion of an additional layer of timber blocks between the CLT slab and the existing joists to increase the floor inertia. The whole system is intended to behave similarly to a Pratt truss, where the inclined screws, ensuring the transfer of shear force between the blocks and the joists, constitute the diagonal truss elements.

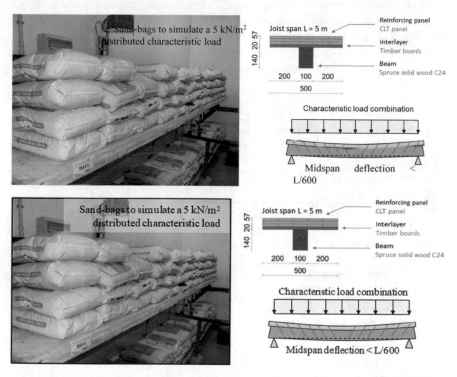

Fig. 14 Example of timber floor specimens retrofitted with CLT panels and pre-stressed by exploiting the screw fastener compression force [97]

5 Conclusion

The analysis of existing retrofitting techniques has shown that a decision on the best method depends on the structure, particularly its age and historical value, its location and the associated seismic risk, availability of space for medium and heavy equipment, and availability of specialized workmanship. Moreover, it is also clear that existing floors cannot be evaluated as isolated elements, but must be analysed considering the structure that supports them and the connections between them and the vertical elements.

A short qualitative overview of the advantages and disadvantages of each method is presented in Table 4. However, it must be made clear that this chapter aims to serve as a starting point and should not be seen as a comprehensive analysis of all available retrofitting methods.

The present chapter has analysed the need for retrofitting of existing timber floors, as well as a range of methods to improve both their in-plane and out-of-plane strength and stiffness. The review shows that this is a very active area of research, and while existing methods can be useful in extending the life of existing structures, it should be expected that new and better methods will be developed in the future.

Table 4 Comparison of the "efficiency" of different retrofitting techniques

Method	In-plane stiffness	Out-of-plane stiffness	Additional weight[b]	Reversibility
Concrete overlay (TCC)	5	5	−5	0
Grid of FRP/metal straps applied to the floor decking	3	0	−1	1
FRP/metal sheets applied to the joist bottom surface	0	4	−1	1
Additional floorboard sheathing	2	1	−2	5
Single plywood/OSB sheet overlay	3	1	−2	5
Thick planks made of solidwood/glulam/LVL running parallel to the joists (TTC)	1	4	−3	4
Slab made of CLT or LVL panels (TTC)[a]	4	5	−3	4

[a]Panel-to-panel edge connection is present
[b]The negative sign is adopted to emphasize the detrimental impact of the additional weight

References

1. Branco JM, Descamps T, Tsakanika E (2018) Repair and strengthening of traditional timber roof and floor structures. Strengthening and retrofitting of existing structures. Springer, Berlin, pp 113–138
2. Parisi MA, Piazza M (2007) Restoration and strengthening of timber structures: principles, criteria, and examples. Pract Periodical Struct Des Constr 12(4):177–185
3. Parisi MA, Piazza M (2015) Seismic strengthening and seismic improvement of timber structures. Constr Build Mater 97:55–66
4. Grubbs A, Hueste M, Bracci JM (2007) Seismic rehabilitation of wood diaphragms in unreinforced masonry buildings. Texas A&M University
5. Binda L, Penazzi D, Valuzzi MR, Cardani G, Baronio G, Modena C (2000) Behavior of historic masonry buildings in seismic areas: lessons learned from the Umbria-Marche earthquake. Proceedings of the 12th international brick/block masonry conference, Madrid, Spain, pp 217–235
6. Spence R, D'Ayala D (1999) Damage assessment and analysis of the 1997 umbria-marche earthquakes. Struct Eng Int 9(3):229–233
7. Dizhur D et al (2011) Performance of masonry buildings and churches in the 22 February 2011 Christchurch earthquake. Bull N Z Soc Earthq Eng 44(4):279–296
8. D'Ayala D, Paganoni S (2011) Assessment and analysis of damage in L'Aquila historic city centre after 6th April 2009. Bull Earthq Eng 9(1):81–104
9. Penna A, Morandi P, Rota M, Manzini CF, da Porto F, Magenes G (2013) Performance of masonry buildings during the Emilia 2012 earthquake. Bull Earthq Eng 12(5):2255–2273
10. Sorrentino L, Cattari S, da Porto F, Magenes G, Penna A (2018) Seismic behaviour of ordinary masonry buildings during the 2016 central Italy earthquakes. Bull Earthq Eng 17:5583–5607
11. Baldessari C, Piazza M, Tomasi R (2009) The refurbishment of existing timber floors: characterization of the in-plane behaviour. Protection of historical buildings: proceedings of the international conference on protection of historical buildings, PROHITECH 09, Rome, Italy, 21–24 June 2009, pp 255–260
12. ATC (Applied Technology Council) (1981) Guidelines for the design of horizontal wood diaphragms. Berkeley, CA: ATC
13. Rizzi E, Giongo I, Ingham J, Dizhur D (2019) Testing and modelling of retrofitted timber diaphragms loaded in-plane. J Struct Eng 146(2)
14. Donà C, De Maria A (2011) (eds) Manuale delle murature storiche. ISBN: 884960403 (in Italian)
15. Circolare 21 gennaio (2019) n.7 C.S.LL.PP, Istruzioni per l'applicazione dell'"Norme tecniche per le costruzioni" di cui al decreto ministeriale 17 gennaio 2018. Ministry of Infrastructures and Transportation, (in Italian)
16. De Benedictis R, De Felice G, Giuffrè A (1993) Anti-seismic renewal of a building. In: Giuffrè (ed) Safety and preservation of historical city centre. Editrice Laterza, Bari (In Italian)
17. Marini A, Giuriani E (2006) Transformation of wooden roof pitches into antiseismic shear resistance diaphragms, structural analysis of historical constructions, New Delhi
18. Parisi MA, Chesi C (2014) Seismic vulnerability of traditional buildings: the effect of roof-masonry walls interaction. NCEE 2014—10th U.S. national conference on earthquake engineering: frontiers of earthquake engineering 2014. Anchorage, USA
19. Senaldi I, Magenes G, Penna A, Galasco A, Rota M (2014) The effect of stiffened floor and roof diaphragms on the experimental seismic response of a full-scale unreinforced stone masonry building. J Earthq Eng 18(3):407–443
20. Brignola A, Pampanin S, Podestà S (2009) Evaluation and control of the in-plane stiffness of timber floors for the performance-based retrofit of URM buildings. Bull NZ Soc Earthq Eng 42(3):204
21. Nakamura H, Derakhshan G, Magenes MC (2017) Griffith, influence of diaphragm flexibility on seismic response of unreinforced masonry buildings. J Earthq Eng 21(6):935–960

22. Scotta R, Trutalli D, Marchi L, Pozza L (2018) Seismic performance of URM buildings with in-plane non stiffened and stiffened timber floors. Eng Struct 167:683–694
23. Spencer RJS, Oliveira CS, D'Ayala D, Papa F, Zuccaro G (2000) The performance of strengthened masonry buildings in recent European earthquakes. In: Proceedings of 12th WCEE conference, Auckland
24. Doglioni F (2000) Handbook (guidelines) for the design of adjustment interventions, seismic strengthening, and renewal of architectural treasures damaged during the Umbria-Marche earthquake in 1997. Official Bulletin of Marche Region, Ancona (in Italian)
25. Hsiao JK, Tezcan J (2012) Seismic retrofitting for chord reinforcement for unreinforced masonry historic buildings with flexible diaphragms. Pract Periodical Struct Design Const ASCE 17(3):102–109
26. Brignola A, Podestà A, Pampanin S (2008) In-plane stiffness of wooden floor. In: Proceedings of the New Zealand society for earthquake engineering conference. University of Canterbury, Canterbury, New Zealand
27. Schiro G, Giongo I, Ingham J, Dizhur D (2018) Lateral performance of as-built and retrofitted timber diaphragm fastener connections. J Mater Civil Eng 30(1)
28. Wilson A, Quenneville PJH, Moon FL, Ingham JM (2014) Lateral performance of nail connections from century old timber floor diaphragms. ASCE J Mater Civil Eng 26(1):202–205
29. European Committee for Standardization (2014) EN 1995-1-1:2004 + A2:2014: Eurocode 5—design of timber structures, Part 1-1, general—common rules and rules for buildings. CEN, Brussels, Belgium
30. ASCE (2017) Seismic evaluation and retrofit of existing buildings. ASCE 41-17. Reston, VA: ASCE
31. ABK–TR–03 (1981) Methodology for mitigation of seismic hazards in existing unreinforced masonry buildings: diaphragm testing, ABK
32. NZSEE (New Zealand Society for Earthquake Engineering) (2017) Assessment and improvement of the structural performance of buildings in earthquakes. NZSEE, Wellington, New Zealand
33. Wilson A, Quenneville PJH, Ingham JM (2014) In-plane orthotropic behaviour of timber floor diaphragms in unreinforced masonry buildings. ASCE J Struct Eng 140(1):04013038
34. Giongo I, Dizhur D, Tomasi R, Ingham JM (2014) Field testing of flexible timber diaphragms in an existing vintage URM building. ASCE J Struct Eng 141(1)
35. Giongo I, Schiro G, Tomasi R, Dizhur D, Ingham J (2015) Seismic assessment procedures for flexible timber diaphragms. In: Historical earthquake-resistant timber framing in the mediterranean area
36. Giongo I, Wilson A, Dizhur DY, Derakhshan H, Tomasi R, Griffith MC, Quenneville P, Ingham J (2014) Detailed seismic assessment and improvement procedure for vintage flexible timber diaphragms. NZSEE Bull 47:97–118
37. Rizzi E, Capovilla M, Giongo I, Piazza M (2017) Numerical study on the in-plane behaviour of existing timber diaphragms strengthened with diagonal sheathing, SHATIS'17. In: International conference on structural health assessment of timber structures, Instanbul, Turkey
38. Rizzi E, Capovilla M, Piazza M, Giongo I (2019) In-plane behaviour of timber diaphragms retrofitted with CLT panels. In: Aguilar R et al (eds) Structural analysis of historical constructions, RILEM Bookseries 18:1613–1622
39. Valluzzi MR, Garbin E, dalla Benetta M, Modena C (2013) Experimental characterization of timber floors strengthened by in-plane improvement techniques. Adv Mater Res 778:682–689
40. Gattesco N, Macorini L (2008) High reversibility technique for in-plane stiffening of wooden floors. Structural analysis of historic construction. Taylor & Francis, London
41. Gattesco N, Macorini L (2014) In-plane stiffening techniques with nail plates or CFRP strips for timber floors in historical masonry buildings. Constr Build Mater 58:64–76
42. Gattesco N, Macorini L (2006) Strengthening and stiffening ancient wooden floors with flat steel profiles. Struct Anal Hist Const, New Delhi

43. Valluzzi MR, Garbin E, Dalla Benetta M, Modena C (2010) In-plane strengthening of timber floors for the seismic improvement of masonry buildings. In: Ceccotti A, Van de Kuilen JW (eds) 11th world conference on timber engineering WCTE 2010, Riva del Garda, Italy, 20–24 June 2010
44. Corradi M, Speranzini E, Borri A, Vignoli A (2006) In-plane shear reinforcement of wood beam floors with FRP. Composites: Part B37:310–319
45. Brignola A, Pampanin S, Podestà S (2012) Experimental evaluation of the in-plane stiffness of timber diaphragms. Earthq Spectra 28(4):1687–1709
46. Giongo I, Dizhur D, Tomasi R, Ingham JM (2013) In plane assessment of existing timber diaphragms in URM buildings via quasi static and dynamic in situ tests. Adv Mater Res 778:495–502
47. Gubana A, Melotto M (2018) Experimental tests on wood-based in-plane strengthening solutions for the seismic retrofit of traditional timber floors. Constr Build Mater 191:290–299
48. Branco JM, Kekeliak M, Lourenço PB (2014) In-plane stiffness of traditional timber floors strengthened with CLT. RILEM Bookseries 9:725–737
49. Schiro G, Rizzi E, Piazza M (2017) Interventions aimed at reducing the excessive deformability of timber floors: strengthening and stiffening techniques according to the new Italian code (NTC). In: Proceedings of the XVII ANIDIS (Italian national association of earthquake engineering) conference, Pistoia, Italy (in Italian)
50. Binda L, Gambarotta L, Lagomarsino S, Modena C (1999) A multilevel approach to the damage assessment and the seismic improvement of masonry buildings in Italy. In: Bernardini A (ed) Seismic damage to masonry buildings, Balkema, Rotterdam
51. Steinberg E, Selle R, Faust T (2003) Connectors for timber-lightweight concrete composite structures. J Struct Eng 129(11):1538–1545
52. Santos P, Martins C, Skinner J, Harris R, Dias A, Godinho L (2015) Modal frequencies of a reinforced timber-concrete composite floor: testing and modeling. J Struct Eng 141(11):04015029
53. Möhler K (1956) Über das tragverhalten von biegeträgern und druckstützen mlt zusammengesetzten querschnitten und nachgiebigen verbindungsmitteln, Habilitation Thesis, TU-Karlsruhe (in German)
54. Frangi A, Fontana M (2003) Elasto-plastic model for timber-concrete composite beams with ductile connection. Struct Eng Int 13(1):47–57
55. Yeoh D, Fragiacomo M, De Franceschi M, Heng Boon K (2011) State of the art on timber-concrete composite structures: literature review. J Struct Eng 137(10):1085–1095
56. Fragiacomo M, Yeoh D (2010) Design of timber-concrete composite beams with notched connections. In: Meeting forty-three of the working commission W18-timber structures, CIB
57. Otero-Chans D, Estévez-Cimadevila J, Suárez-Riestra F, Martín-Gutiérrez E (2018) Experimental analysis of glued-in steel plates used as shear connectors in timber-concrete composites. Eng Struct 170:0141-0296
58. Van der Linden M (1999) Timber–concrete composite floors. PhD thesis, Delft University of Technology
59. Dias A, Jorge F (2011) The effect of ductile connectors on the behaviour of timber–concrete composite beams. Eng Struct 33:3033–3042
60. Sebastian WM, Piazza M, Harvey T, Webster T (2018) Forward and Reverse shear transfer in beech LVL-concrete composites with singly inclined coach screw connectors. Eng Struct 175:231–244
61. Sebastian WM, Mudie J, Cox G, Piazza M, Tomasi R, Giongo I (2016) Insight into mechanics of externally indeterminate hardwood-concrete composite beams. Constr Build Mater 102(2):1029–1048
62. Ceccotti A (2002) Composite concrete-timber structures. Prog Struct Mat Eng 4(3):264–275
63. Fragiacomo M (2006) Long-term behavior of timber–concrete composite beams. II: Numerical analysis and simplified evaluation. J Struct Eng 132(1):23–33
64. Schänzlin J (2015) Zum Langzeitverhalten von Brettstapel-Beton-Verbunddecken. PhD thesis, Institut für Konstruktion und Entwurf, University of Stuttgart

65. Dias A, Schänzlin J, Dietsch P (eds) (2018) Design on timber-concrete composite structures—a state-of-the-art report by COST action FP1402/WG 4. Shaker Verlag Aachen
66. Dias A, Skinner J, Crews K, Tannert T (2016) Timber-concrete-composites increasing the use of timber in construction. Eur J Wood Wood Prod 74(3):443–451
67. Pizzo B, Smedley D (2015) Adhesives for on-site bonding: characteristics, testing and prospects. Constr Build Mater 97:67–77
68. Raftery G, Harte A, Rodd P (2009) Bonding of FRP materials to wood using thin epoxy gluelines. Int J Adhes Adhes 29:580–588
69. Schober K, Harte AM, Kliger R, Jockwer R, Xu Q, Chen J-F (2015) FRP reinforcement of timber structures, construction and building materials. Special Issue Reinforcement Timber Struct 97:106–118
70. Hollaway L, Teng J (2008) Strengthening and rehabilitation of civil infrastructures using fibre-reinforced polymer (FRP) composites. Woodhead Publishing Limited, Cambridge, UK
71. Greco A, Maffezzoli A, Casciaro G, Caretto F (2014) Mechanical properties of basalt fibers and their adhesion to polypropylene matrices. Compos B Eng 67:233–238
72. Nowak TP, Jasienko J, Czepizak D (2013) Experimental tests and numerical analysis of historic bent timber elements reinforced with CFRP strips. Constr Build Mater 40:197–206
73. Kliger R, Johansson M, Crocetti R (2008) Strengthening timber with CFRP or steel plates—short and long-term performance. In: Proceedings of world conference on timber engineering, Miyazaki, Japan
74. Borri A, Corradi M, Grazini A (2005) A method for flexural reinforcement of old wood beams with CFRP materials. Compos B Eng 36:143–153
75. Li YF, Xie YM, Tsai MJ (2009) Enhancement of the flexural performance of retrofitted wood beams using CFRP composite sheets. Const Build Mater 23:411–422
76. Johns KC, Lacroix S (2000) Composite reinforcement of timber in bending. Can J Civ Eng 27:899–906
77. Valluzzi MR, Nardon F, Garbin E, Panizza M (2016) Multi-scale characterization of moisture and thermal cycle effects on composite-to-timber strengthening. Const Build Mater 102:1070–1083
78. Hoseinpour H, Valluzzi MR, Garbin E, Panizza M (2018) Analytical investigation of timber beams strengthened with composite materials. Const Build Mater 191:1242–1251
79. CNR-DT 201/2005 (2005) Instructions for strengthening interventions on timber structures by using fiber reinforced composites (in Italian)
80. CNR-DT 200/2004 (2004) Guide for the design and construction of externally bonded FRP systems for strengthening existing structures: materials, RC structures, prestressed RC structures, masonry structures
81. Martin A, Tingley D (2000) Fire resistance of FRP reinforced glulam beams. In: World conference on timber engineering, Whistler Resort, British Columbia, Canada
82. Haghani R, Al-Emrani M (2014) A new method and device for application of bonded prestressed FRP laminates. In: Proceedings of the second international conference on advances in civil and structural engineering, Kuala Lampur, Malaysia
83. Negrão J, Brunner M, Lehmann M (2008) Pre-stressing of timber. In: Richter K, Cruz H (eds) COST E34—bonding of wood—WG1: bonding on site—core document
84. Kliger IR, Haghani R, Brunner M, Harte AM, Schober K (2016) Wood-based beams strengthened with FRP laminates: improved performance with pre-stressed systems. Eur J Wood Wood Prod 74(3):319–330
85. Piazza M (1994) Restoration of timber floors via a composite timber-timber solution. In: Proceedings of the technical workshop RILEM timber: a structural material from the past to the future
86. Modena C, Valluzzi MR, Garbin E, da Porto F (2004) A strengthening technique for timber floors using traditional materials. In: Proceedings of the fourth international seminar on structural analysis of historical constructions, Padova, Italy, 10–13 Nov 2004, pp 911–921
87. Valluzzi MR, Garbin E, Modena C (2007) Flexural strengthening of timber beams by traditional and innovative techniques. J Build Appraisal 3(2):125–143

88. Piazza M, Riggio M, Tomasi R, Giongo I (2010) Comparison of in situ and laboratory testing for the characterization of old timber beams before and after intervention. Adv Mater Res 133–134:1101–1106
89. Giongo I, Piazza M, Tomasi R (2012) Out of plane refurbishment techniques of existing timber floors by means of timber to timber composite structures. In: WCTE, world conference on timber engineering, Auckland, New Zealand
90. Riggio M, Tomasi R, Piazza M (2014) Refurbishment of a traditional timber floor with a reversible technique: Importance of the investigation campaign for design and control of the intervention. Int J Architectural Heritage 8(1):74–93
91. Bejtka I, Blass HJ (2002) Joints with inclined screws. In: Proceedings from meeting thirty-five of the international council for building research studies and documentation, CIB, Working Commission W18—timber structure, Kyoto, Japan
92. Tomasi R, Crosatti A, Piazza M (2010) Theoretical and experimental analysis of timber-to-timber joints connected with inclined screws. Constr Build Mater 24:1560–1571
93. Schiro G, Giongo I, Sebastian W, Riccadonna D, Piazza M (2018) Testing of timber screw-connections in hybrid configurations. Constr Build Mater 171:170–186
94. Giongo I, Piazza M, Tomasi R (2013) Investigation on the self tapping screws capability to induce internal stress in timber elements. Adv Mater Res 778:604–611
95. Giongo I, Schiro G, Riccadonna D (2019) Innovative pre-stressing and cambering of timber-to-timber composite beams. Compos Struct, p 226
96. Giongo I, Schiro G, Walsh K, Riccadonna D (2019) Experimental testing of pre-stressed timber-to-timber composite (TTC) floors. Eng Struct 201:109808
97. Riccadonna D, Walsh K, Schiro G, Piazza M, Giongo I (2020) Testing of long-term behaviour of pre-stressed timber-to-timber composite (TTC) floors. Const Build Mater 236
98. de Lima L, Costa AA, Rodrigues CF (2018) On the use of prestress for structural strengthening of timber beams: assessment with numerical support and experimental validation. Int J Arch Heritage 12(4):710–725
99. Gubana A (2010) Experimental tests on timber-to-cross lam composite section beams. In: Proceedings of 11th world conference on timber engineering WCTE 2010, Riva del Garda, TN, Italy
100. Hossain A, Popovski M, Tannert T (2018) Cross-laminated timber connections assembled with a combination of screws in withdrawal and screws in shear. Eng Struct 168:1–11
101. Sullivan K, Miller TH, Gupta R (2018) Behavior of cross-laminated timber diaphragm connections with self-tapping. Eng Struct 168:505–524
102. Roensmaens B, Van Parys L, Carpentier O, Descamps T (2018) Refurbishment of existing timber floors with screwed CLT panels. Int J Architectural Heritage 12(4):622–631
103. Roensmaens B, Van Parys L, Branco J, Descamps T (2019) Proposal of a CLT reinforcement of old timber floors. In: Aguilar R et al (2019) Structural analysis of historical constructions. RILEM Bookseries, p 18

Reinforcement of Traditional Carpentry Joints

Jorge M. Branco, Maxime Verbist, and Eleftheria Tsakanika

Abstract This chapter focuses on the reinforcement of traditional timber carpentry joints according to different standards, recommendations, case studies, as well as analytical, numerical and experimental research works. The aim of the review is to present the state-of-the-art methodologies that can be at the disposal of carpenters, architects and engineers. In order to understand how traditional carpentry joints, work and their failure mechanisms, such as compressive crushing, shear and tensile cracks, six geometrical typologies of such joints are discussed. Finally, for the above typologies, several reinforcement strategies are presented by defining their objectives, methodologies, traditional and contemporary techniques, performance criteria and applicability areas. When assessing and reinforcing traditional carpentry joints, some challenges may come up, namely their design, based on the ratios of stiffness and load-bearing capacities. These challenges should thus be a focal point for further research in the near future.

1 Introduction

Wood can be easily found in everyday life all around the world. For millennia, humans have widely used it as a natural construction material. The preservation of timber structures has been of utmost importance for some decades now in many European countries. The debate concerns mostly existing buildings, especially buildings of historical and cultural importance. On the other hand, their preservation is often threatened due to lack of earlier structural maintenance and monitoring. Under

J. M. Branco (✉)
Department of Civil Engineering, ISISE, University of Minho, Guimarães, Portugal
e-mail: jbranco@civil.uminho.pt

M. Verbist
Civil Engineering Department, ISISE, University of Minho, Braga, Portugal

E. Tsakanika
School of Architecture, National Technical University of Athens, Athens, Greece

© RILEM 2021

J. Branco et al. (eds.), *Reinforcement of Timber Elements in Existing Structures*,
RILEM State-of-the-Art Reports 33,
https://doi.org/10.1007/978-3-030-67794-7_12

247

unsuitable environmental conditions, timber structures may suffer from wood deterioration triggered by biotic, chemical and physical agents, resulting in problems at different levels, concerning their safety and long-term mechanical performance. Simultaneously, service or loading and structural modifications, occur often in the lifetime of existing buildings as they go through successive restoration or rehabilitation stages. For timber structures characterized by initial poor design or by poor previous interventions, their stiffness and load-bearing capacity may no longer meet the requirements, even for their original or normal use, causing failures or unacceptable deformation. Moreover, several existing buildings may change use and heavier loads need to be carried by the timber load bearing structures. Therefore, structural reinforcement becomes a necessity nowadays to ensure the preservation of existing buildings over time.

In timber structures (e.g. roofs, walls, floors, etc.), it is quite easy to design members in compression, shear, tension, bending and torsion with respect to their cross-sections and axial lengths based on standardized recommendations, while considering some instability phenomena (i.e. buckling and warping). For traditional carpentry joints, the loading transfer and related stress-strain distributions in-between adjoining timber pieces are more complex to determine, since no conventional rules exist so far. Designing such timber-to-timber connections, based on ratios of stiffness and load-bearing capacities is thus very challenging. Reinforcement of traditional carpentry joints should also be considered, for seismic actions too, in order to reduce the uncertainty level on their respective mechanical performance, resulting in a safer design for the existing timber structures, increasing this way the possibility to be preserved.

In the past, carpentry joints were not designed to transfer loads through shear via metal fasteners such as nails, screws or bolts, all of which belong to the so-called dowel-type connections. In fact, their ability to carry these loads was achieved through friction and via the shear capacity on the contact surfaces of the connection increased by significant compression at an angle to the grain. For these joints, direct contact was achieved from special cuttings and notches curved at the timber connected members. That is why they are today called "traditional carpentry joints". Nonetheless, the few metal fasteners that were used in parallel with traditional carpentry joints, did not carry the loads. They ensured, by tightening, a good fitting and contact of the members in the region of the connection.

Nowadays, the contemporary reinforcement methods of traditional carpentry joints can be achieved by reproducing the traditional techniques undertaking the preservation of either the original metal fasteners or their substitution including new connectors. Innovative techniques have been made possible today too, thanks to the emergence of advanced technologies in the field of construction and engineering.

All these reinforcement techniques have to fulfil two main objectives: (i) they must ensure tight contact of connected timber members for an efficient load transfer; (ii) they must prevent dismantling of the joint under reverse loads from severe wind, snow or seismic events. Some of the most important parameters for the choice of the reinforcement technique suitable for each case, are the loading condition of the joint and the type of internal forces (i.e. tension, compression, shear, and/or bending moment), transferred between the connected timber members. Each technique has advantages, disadvantages and direct consequences regarding the mechanical performance of the reinforced joints (i.e. strength, stiffness and failure mechanisms).

Based on the above, this chapter focuses on the reinforcement of traditional timber carpentry joints according to different standards, recommendations, case studies, as well as analytical, numerical and experimental research works. The state-of-the-art methods that are presented and can be at the disposal of carpenters, architects and engineers have been approached in two steps: (i) analysis of the main characteristics of the traditional carpentry joints; (ii) reinforcement methods. It is hoped that this documentation would assist practitioners by drawing their attention to some reinforcement methods and techniques that can be adopted for traditional carpentry joints in existing timber structures, and by increasing their awareness of the conditions that potentially will threaten the service life of the joints.

2   Traditional Carpentry Joints

Before presenting any intervening methods in existing buildings, it is crucial to understand the different typologies and geometries of traditional carpentry joints commonly used in timber structures, the types of internal forces the joints can bear, and the failure modes likely to occur. Drawing the attention on available standards, experimental and numerical studies, the following paragraphs deal with the taxonomy, geometry, and failure mechanisms of traditional carpentry joints.

2.1   Taxonomy and Geometry

From the literature [1–5], an exhaustive taxonomy of traditional timber carpentry joints can be established with respect to their geometrical configurations. Six typologies of joints are presented (Fig. 1): (a) lap joints; (b) tabled joints; (c) scarf joints; (d) heading joints; (e) mortise-tenon joints; (f) step joints. All these connections are designed to transfer mainly the compressive loads in traditional timber structures (e.g. roofs, walls, floors, etc.). Due to their specific geometry, some of them are also designed to be able to transfer low to moderate shear and tensile forces. Information concerning experimental and numerical works on the mechanical characterization of traditional carpentry joints with respect to the six geometrical typologies of connections are described below.

Lap joints [6–9], are assembled by two timber pieces, overlapped either end-to-end along their lengths or at an angle forming a cross, L or T-type configuration. For the simple half-lap splice joint (Fig. 1a), each timber member is split into two equal halves along its thickness, so that the thickness of the resulting connection is equal to the thickness of one of the members. Although they usually work in compression, half-lap joints can be either tenoned or dovetailed, with the objective

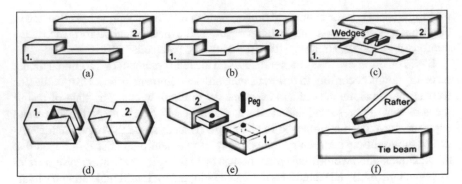

Fig. 1 Illustrations of the most common traditional carpentry joints: **a** simple half-lap splice joint; **b** halved and tabled splice joint; **c** scarf joint with under-squinted ends and wedges (Trait-de-Jupiter); **d** dovetailed heading joint; **e** mortise-tenon joint with peg; **f** single step joint

to gain some flexural behaviour or axial tensile load bearing capacity respectively. If half-lap joints are also subjected to transverse tensile loads, they should thus be reinforced to prevent out-of-plane displacement.

Tabled joints [8, 10–13], are used to increase the material length by attaching two timber pieces end-to-end, thanks to the presence of flanges from either side of the overlap surface. The halved and tabled splice joint (Fig. 1b) can bear compressive, tensile and shear loads in the axial direction of the members. Furthermore, tabled joints can be dovetailed or tenoned at their flanges, to provide rotational stiffness and bending strength. Reinforcement could also be used to enhance the mechanical performance of such connections.

A disadvantage of the above connections is the reduction of the member section at the joint area.

In cases where the timbers are not available in the required length, scarf joints [6, 8, 10] enable as well, the end-to-end connection of two timber members through a sloping cut of overlapped surfaces. The scarf joint in beams with under-squinted ends and wedges, "Trait-de-Jupiter" in French, due to its resemblance to lightning (Fig. 1c), mostly carries compression but it can carry tension and shear along the axial direction of the members. When pressing adjoining timber pieces in axial and transversal directions, significant frictional forces appear on the overlapped surfaces, leading to an increase in the mechanical performance of scarf joints. Wedges are used in some cases in order to ensure the tight fitting of the connected members. Nonetheless, reinforcement could be implemented in such connections to safeguard their respective tensile and flexural behaviour, if needed.

Heading joints [13–15], result from joining together two timber elements end-to-end along their lengths; they mostly counteract axial compression (parallel to the grain) and axial tension, only when dovetailed (Fig. 1d), or tenoned with a peg. In case they do not feature any special geometrical shaping at their ends, simple heading joints should be reinforced to hold in place adjoining timber elements subjected to shear, tension or transverse compression.

A mortise-tenon joint [16–19], comprises two main components: (i) the tenon, which is a projection made on one end of the first timber element; (ii) the mortise, which is an opening in the second timber element intended to receive the tenon. Both mortise and tenon components have usually round, square or rectangular shape. In addition to the significant compressive strength of the butt contact surfaces, mortise-tenon joints feature some rotational stiffness and flexural capacity. Pegs, wedges or tusks can be inserted in such connections (Fig. 1e), subjected to moderate tensile loads, to avoid axial and transverse displacements of the timber members.

Step joints [20–24], are oblique-type connections with V-shaped groove; they mostly work in compression at the contact surfaces and in tension and shear along the grain at the end of the notched timber member. In a king-post truss, a single step joint (Fig. 1f) is commonly used to link the rafter with the tie beam, the strut to the king-post (Fig. 6, right), and the rafters to the king-post. The stiffness and load-bearing capacity of step joints are strongly influenced by the geometry of the heel. The heel can be duplicated, moved at the back side of the connection, or even tenoned, when the length from the notch till the end of the tie-beam (heel depth) is not enough to carry the shear forces, or out-of-plane displacements need to be reduced. In seismic loading conditions, step joints could be reinforced, if needed, to enhance their bending and tensile performance.

2.2 Failure Mechanisms

Eurocode 5 [25], provides standardized recommendations to design timber members in compression, shear, tension, bending, and torsion with respect to their cross-section and axial length, while considering some instability phenomena (i.e. buckling and warping). Nonetheless, little information can be found, even in National Annexes, on how to design traditional carpentry connections based on their geometry with the objective to prevent the occurrence of failure modes in adjoining members. Compressive crushing, shear cracks and tensile cracks have been emphasized in experimental investigations [8, 10, 13, 14, 22], as the three main failure mechanisms (Fig. 2), likely to occur in the following four types of traditional carpentry joints: (a) dovetailed heading joint; (b) half-lap splice joint with under-squinted ends; (c) halved and tabled splice joint; (d) single step joint.

Since the dovetailed heading joint works in axial tension (Fig. 2a), the shear and tensile cracks may occur parallel and perpendicular to the grain respectively in the dovetailed tenon, when the tenon is held in place inside a stiff mortise [14]. In other cases, the shoulders at the mortise-side tend to open, causing tensile cracks along the grain while the tenon-member moves [13].

When this traditional carpentry joint bears compressive loads in the grain direction, crushing may occur at the bottom surface of the dovetailed mortise-tenon and in the shoulders.

In axial compression, the half-lap splice joint with under-squinted ends (Fig. 2b) can fail, either by shear-cracking at the notch depth along the grain or by crushing at

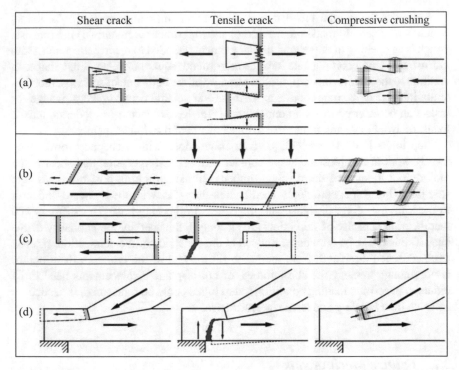

Fig. 2 Failure mechanisms in traditional carpentry joints: **a** dovetailed heading joint; **b** half-lap splice joint with under-squinted ends; **c** halved and tabled splice joint; **d** single step joint

the front-notch surface [8]. Under bending loads, the connection tends to open further with the emergence of tensile cracks along the grain at the notch depth. Regarding the halved and tabled splice joint in axial tension (Fig. 2c), the shear crack occurs along the grain at the flange depth while the crushing takes place at the flange edge in both timber members [8, 10, 13]. In the case where the flange depth is too high, tensile cracks perpendicular to the grain may emerge in one or both adjoining elements.

When a single step joint bears axial compression in the rafter (Fig. 2d), a shear crack occurs in the tie beam-end along the grain at the heel depth while crushing is more concentrated at the front-notch surfaces for low rafter skew angles [22]. In addition, a mix of shear and tensile cracks may occur in the tie beam-end as a result of excessive bending moment and vertical shear forces in the timber member when there is a significant distance between the joint heel and the support area. Lastly, the wood compressive strength governs the occurrence of crushing at the contact surfaces of the connection; it however reduces with an increase in the inclination angle of the loaded area to the grain, for which high variations of this mechanical property can be noticed between 0 and 30° [26, 27].

In wooden notched pieces constituted of any of the major traditional carpentry joints (i.e. lap, tabled, scarf or step joints), a non-uniform shear stress distribution leads to shear crack development over some length parallel to the grain at the notch depth [28, 29]. The shear stress along the grain may often be paired with tensile or compressive stress distribution perpendicular to the grain and highly concentrated in the vicinity of the notch. In conformity with Amendment 1 of Eurocode 5 [30], the presence of longitudinal cracks in timber adjoining members subject to wetting-drying cycles or bending, may trigger a reduction in the shear strength along the grain. Some studies on the development, modelling and effect of cracks on the wood mechanical performance can be found in [31, 32].

For traditional carpentry joints that have been pegged or fastened to carry tensile forces (i.e. lap joints, tabled joints, scarf joints and mortise-tenon joints), their failure modes are very similar to those that occur in standardized dowel-type joints working in shear [6, 7, 9, 16]. In conformity with Eurocode 5 [25], two main categories of ductile failure modes occur in timber-to-timber connections reinforced with fasteners per single or double shear planes (Fig. 3): (i) crushing of timber in the vicinity of the fasteners, also called embedding or embedment; (ii) plastic hinge of fasteners. To achieve an optimal design of such connections, the emergence of plastic hinges in the steel fastener in every timber member should be preferred.

Ductility is of major importance, especially in seismic design [33], for the redistribution of internal forces in non-statically determinate structures or for the improvement of the robustness of a structure. Nonetheless, the governance of one failure mode among others is highly conditioned by the geometry and mechanical properties of the joint components. Moreover, many fracture forms (Fig. 4) may still appear, this time as brittle failure modes, due to high combination of shear stress parallel to the

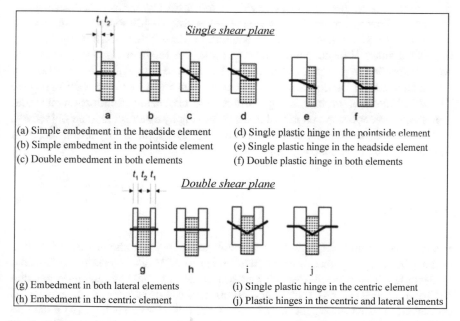

(a) Simple embedment in the headside element   (d) Single plastic hinge in the pointside element
(b) Simple embedment in the pointside element   (e) Single plastic hinge in the headside element
(c) Double embedment in both elements   (f) Double plastic hinge in both elements

(g) Embedment in both lateral elements   (i) Single plastic hinge in the centric element
(h) Embedment in the centric element   (j) Plastic hinges in the centric and lateral elements

Fig. 3 Failure mechanisms in standardized dowel-type joints, namely timber-to-timber connections with fasteners in single and double shear planes. Adapted from Eurocode 5 [25]

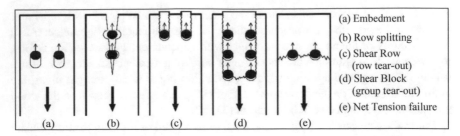

Fig. 4 Embedment versus brittle failure modes in standardized dowel-type joints. Adapted from [34]

grain and tensile stress perpendicular to the grain, both concentrated in the vicinity of the fasteners [34]. A common, brittle and severe damage in timber connections, is splitting due to non-appropriate end distances. These fracture forms should carefully be prevented by checking the minimal distance and spacing recommendations, used for each fastener (e.g. nails, self-tapping screws, bolts, dowels etc.), and the minimal thickness of adjoining elements [25, 35, 36].

3 Reinforcement Methods

After assessing timber structures based on the most appropriate methodology of inspection, documentation, diagnosis and assessment [37–40], some decisions should be taken to solve problems noticed in situ and to prevent them from intensifying or happening again. Whether a minor or major problem, the interventions have to fulfil some restoration or preservation principles beforehand [41–44], so that the function and authenticity of the timber structures can be conserved for the remaining service life of the buildings, while ensuring a sufficient level of safety. Reinforcement strategies and interventions are presented in the following by establishing their objectives, methodologies, techniques, performance criteria and applicability areas, based on in situ case studies and lab experiments.

3.1 Objectives

Intervention methods focusing on traditional carpentry joints should be promoted in existing buildings to protect timber structures from failing or at worst from collapsing. In accordance with the restoration and preservation principles regarding existing timber structures [41–44], such interventions have to fulfil three main objectives: (i) conservation or enhancement of both structural integrity and safety; (ii) reduction of the risk of brittle failures (e.g. shear and tensile cracks, timber splitting), while promoting ductile behaviour (e.g. tensile yielding or plastic hinge in steel fasteners, crushing of fibres); (iii) restoration or improvement of the mechanical performance in terms of stiffness and strength (e.g. seismic behaviour, cycling loading, excessive deflection). On the above basis, four major intervention strategies are emphasized here: (i) reinforcement; (ii) strengthening; (iii) retrofitting; (iv) repair.

Reinforcement [43–47] aims to upgrade damaged and undamaged structures in order to enhance their initial mechanical behaviour. On the other hand, the term "strengthening" is used to define the upgrading process of structures that already had the benefit of some interventions in the past. Retrofitting focuses on stabilizing degraded structures structurally, with greater emphasis on preservation and authenticity issues. The load-bearing capacity and stiffness of reinforced traditional timber carpentry joints should be higher than the initial values. New mechanical performance (e.g. ductility in shear, tension) can also be obtained after reinforcements that overcome weaknesses of the wood material, such as natural defects (e.g. drying cracks, knots, grain deflection, etc.), biological and chemical wood-deterioration, brittle failure modes, and low strength in compression and tension perpendicular to the grain and in shear parallel to the grain.

In contrast to the other three intervention strategies, repair [44, 45, 47], ideally results in regaining the initial mechanical behaviour of the structure (i.e. stiffness, strength, and failure modes), before damage occurred. Repair, in many cases laborious and onerous, is usually promoted for traditional carpentry joints highly degraded due to biotic (e.g. wood-destroying fungi and insects), physical or chemical agents (e.g. fire combustion). Common repair techniques, such as the use of prostheses [42, 48, 49], aim at substituting the damaged timber part by new members or joints made of wood or other materials (e.g. resin), which are usually connected to the existing undamaged timber members using lap and other type of joints, fastened and/or glued with resin and different types of rods.

For earthquake-resistant interventions, special attention should be paid to reinforced traditional carpentry joints [50–54], by: (i) protecting them from brittle failures; (ii) providing them as much as possible with post-elastic behaviour and ductility while keeping the timber members in an elastic state. Since wood is a brittle material in tension and shear, with limited post-elastic resources in compression, timber connections become the only possible location of post-elasticity yielding and energy dissipation in timber structures. This is why the emergence of brittle failure modes should be avoided. Nevertheless, timber splitting along the grain may still occur due to high tensile stresses perpendicular to the grain concentrated in the vicinity of the reinforced joint area, resulting in a negative effect on the structural integrity. Thereby, timber splitting, as mentioned before, should be taken into account at the first stage of the reinforcement design of timber connections subjected to axial and transverse loads.

3.2 Methodologies and Techniques

For a better understanding of how to efficiently reinforce traditional carpentry joints, several methodologies and techniques have been illustrated (Fig. 5; Table 1), for the simplest geometrical configuration of step joints, namely the single step joint. The reinforcement methodologies and techniques presented here could also be applied to lap joints, tabled joints, scarf joints, heading joints, mortise-tenon joints, and the other step joints, since all these connections share similar failure mechanisms (e.g. crushing and tensile and shear cracks).

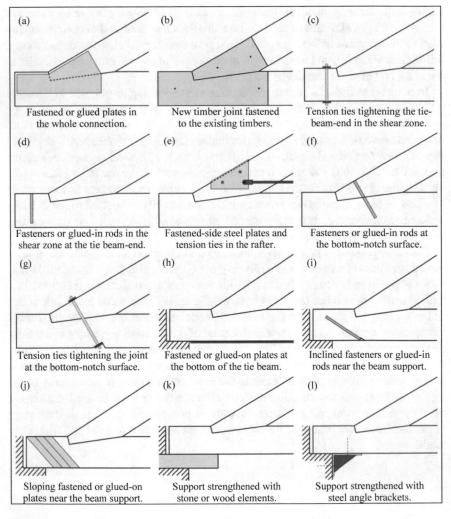

Fig. 5 Illustration of reinforcement techniques for the single step joint

Table 1 Overview of reinforcement techniques for the single step joint classified into five preventive methodologies and three categories of intervention techniques

Reinforcement techniques	Load redistribution	Compressive crushing	Shear crack	Bending	Vertical shear
Figure 5a[2,3]	X	X	X		
Figure 5b[3]	X	X	X		
Figure 5c[3]			X		
Figure 5d[1,2]			X		
Figure 5e[3]	X				
Figure 5f[1,2]	X				
Figure 5g[3]	X				
Figure 5h[3]				X	
Figure 5i[1,2]					X
Figure 5j[3]					X
Figure 5k[3]				X	X
Figure 5l[3]				X	X

Legend 1—dowel-type; 2—inserts; 3—elements set; X—preventive methodology checked

All the reinforcement techniques can be classified into three groups: (i) dowel-type; (ii) inserts; (iii) elements set. The term "dowel-type" is used to describe timber-to-timber connections reinforced with metal fasteners (e.g. dowel, nails, bolts, screws) or wooden components (e.g. pegs, wedges, tusks). The "inserts" group comprises elements (e.g. rods, bars, plates) connected or glued inside adjoining timber pieces. The material used for this reinforcement techniques can be wood, steel, reinforced concrete, glass or carbon fibres, and epoxy resin. Finally, the "elements set" group is made up of external members (e.g. wires, bars, plates, brackets) that are either glued with adhesives or connected with metal fasteners to adjoining timber pieces. This group shares the same materials used for the "inserts" reinforcement techniques.

As detailed in Table 1, all the reinforcement possibilities illustrated for the single step joint (Fig. 5) have been sorted into the three groups of techniques already defined, and into five established methodologies, as per the type of failure mode and the load transfer occurring in the respective timber adjoining members.

- As a means of prevention of compressive crushing at the contact surfaces in the rafter and tie beam (Fig. 2d), the first reinforcement methodology aims at enhancing the compressive capacity of the joint. To this end, fastened or glued plates should be implemented in the whole connection (Fig. 5a), or new timber elements should be fastened to the existing ones by replicating the traditional geometry of the single step joint (Fig. 5b), which would increase the cross-section of adjoining timber members;
- In case of significant rafter thrust, shear cracks may appear along the grain at the heel depth in the tie beam-end (Fig. 2d). To avoid this brittle failure mode, the

second reinforcement methodology aims at increasing the shear capacity related to the tie beam-end. For this purpose, two of the techniques previously stated should be applied (Fig. 5a, b), and more specifically, the tie beam-end in the shear zone should be tightened by external tension rods or ties (Fig. 5c), placed transversally to the grain. In a similar way, fasteners or glued-in rods should also be implemented in the shear zone at the tie beam-end (Fig. 5d);

- The third reinforcement methodology aims at reducing compressive-shear strain, which is mostly concentrated in the joint heel and triggers the emergence of shear crack in the tie beam-end and crushing at the front-notch surface (Fig. 2d). Force redistribution in the rafters should be ensured using two of the techniques previously stated (Fig. 5a, b), or external systems relying on other structural members, such as steel plates fastened to the side of both rafters connected to tension cables or rods (Figs. 5e and 8). Another solution is to partly transfer the rafter forces at the bottom-notch surface through fasteners or glued-in rods (Fig. 5f) and external tension ties (Fig. 5g);

- As a result of excessive bending moment and vertical shear forces in the tie beam-end over the distance between the joint heel and the support area, shear-tensile cracks may also occur, leading to brittle failure modes (Fig. 2d). To overcome this, the last two reinforcement methodologies can be proposed: (i) enhancing the bending and, thus, the shear-tensile strength along the grain in the inferior part of the tie beam-end over the distance between the joint heel and support area; (ii) improving the resistance to vertical shear forces near the support area by providing higher shear-tensile strength perpendicular to the grain in the tie-beam cross-section. For the first strategy, fastened or glued-on plates (Fig. 5h) should be implemented at the bottom of the tie beam, while, for the second, inclined fasteners, glued-in rods (Fig. 5i) or glued-side plates (Fig. 5j) should be preferred near the tie-beam support. Finally, stone, wood (Fig. 5k) or steel pieces (Fig. 5l), placed under the tie-beams can increase and strengthen the support area of the timber roof, and can significantly reduce the bending moment and vertical shear forces at the tie-beam. These elements maybe either single pieces placed only under each tie-beam, or longitudinal elements placed under the tie-beams along the upper part of the supporting walls.

3.3 Case Studies

In the previous section, some methodologies and intervention techniques were defined and illustrated, with greater focus on the reinforcement of a certain type of traditional carpentry joint (i.e. single step joint) and highlighting the related failure mechanisms and loading distribution. As additional information, in situ case studies and lab experimentations, which encompass different typologies of carpentry joints reinforced with traditional and contemporary techniques, are introduced in this section.

In existing timber structures, it is crucial to permanently keep all carpentry joint surfaces in tight contact to ensure sufficient strength and stiffness and, thus, efficient load transfer from one member to another. In case of reverse loads or uplift (e.g. snow, wind and seismic events), initial poor construction, or high shrinkage of the wood material, carpentry joints may develop gaps in-between assembly members, leading to the reduction of their contact areas and, thus, to the reduction of their mechanical performance. To solve this problem, traditional reinforcement methods consisted in placing wooden wedges or inserts to ensure perfect contact of adjoining timber pieces (Fig. 6a). Screws or nails were often used to keep the wedges and inserts in place (Fig. 6b). Experimentations performed on single step joints with mortise and tenon [51], showed that implementing a wooden wedge to fill a gap at the front-notch surface helped restore the rotational behaviour of the original carpentry joint.

Traditional and contemporary reinforcement techniques for the single step joint can be: stirrups connected to timber members by dowel-type fasteners (nails, bolts, screws, etc.) (Fig. 7a), Internal bolts, dowels, or screws placed vertically (Fig. 7b), binding strips (Fig. 7c), bolted or screwed external steel plates and external tension clamps (Fig. 7d). In the past, all of these techniques were often used in historic roofs, either at the early design stage of the joints or added later during the construction phase to overcome the mechanical weaknesses and lack of wood-cutting accuracy at the contact surfaces of the joint [46, 50–52].

Although some carpentry connections still feature some moment-resisting capacity in case of reverse loads (e.g. wind and seismic events), all the reinforcement techniques previously stated, especially the metal stirrups (Fig. 7a) and internal bolts (Fig. 7b), result in a significant increase in the equivalent viscous damping ratio, higher strength and significant improvement in the joint ductility [50, 51] (Fig. 8).

In most cases, the reinforcement of step joints consists in implementing metal fasteners (e.g. nails, screws, bolts) or steel plates to prevent the emergence of shear

Wooden wedges at the front-notch surface.

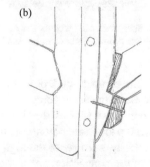
Wooden wedges fastened with nails or screws at notch area.

Fig. 6 Traditional reinforcement techniques for single step joints, that keep adjoining pieces in contact, in order compression forces to be transferred

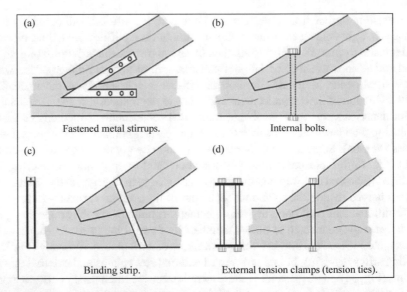

Fig. 7 Contemporary reinforcement of the single step joint reproducing traditional techniques [46]

Fig. 8 Timber trusses in Queen's Tower Estate Ilion, Athens, Greece. Joints reinforced with dowel-type fasteners (bolts), timber inserts, steel rods and external steel plates

cracks in the tie beam-end, or to reinforce a beam-end that does not have the required length to carry the shear forces (Fig. 9a) [46, 47, 55, 56].

Located at the foot of the timber trusses, and in contact with occasionally wet masonry, step joints are potentially problematic connections with regard to their long-term durability, because of the high risk of wood deterioration (i.e. fungal rot, insect attacks) in the connection zone. Most of the time, advanced biological degradation can be spotted in the beam-end, featuring very low residual mechanical performance in compression and shear, which thus leads to a significant need for repair. In that case, the use of a wooden prosthesis to substitute the deteriorated part of the tie beam-end and the reinforcement of the new connection to the existing timber member with

(a) Tie beam-end reinforced with screws.

(b) $N, \cos \alpha$
Screwed timber prosthesis at the tie beam-end.

(c) Wooden insert and metal binding strip.

(d) External steel angled tension ties.

Fig. 9 Contemporary reinforcement techniques for step joints [46, 57]

metal fasteners, such as screws (Fig. 9b) or steel plates, could be used [46, 47].[1] If the prosthesis is found inefficient for the prevention of wood-degradation, either wooden inserts combined with metal binding strips (Fig. 9c) or external steel angled tension ties (Fig. 9d) could be used, to counteract the horizontal thrust transferred from the rafter to the tie beam-end [57].

For the dovetailed half-lap joints (Fig. 10a) [46], a wooden dowel (also called peg or pin) was traditionally implemented at their initial design stage to keep adjoining pieces together and, thus, prevent any gap in between timbers. Furthermore, the wooden peg can also be used to bear low to moderate tension forces. Nonetheless, timber splitting along the grain may occur as a brittle failure mode in the vicinity of the peg, due to moisture changes, small edge distances, or high concentration of shear-tensile stress in the dovetailed member. In that case, the traditional reinforcement of such joints consists in using metal fasteners (e.g. bolts, nails, screws) in the vicinity of the existing peg (Fig. 10b), to enhance the stiffness and shear-tensile load-bearing capacity of the dowel-type connection. The edge distances of the new fasteners must be taken into account too.

In the case that the wooden pins are broken, two external reinforcement techniques working in tension [46], could be considered: (i) metal straps (Fig. 11a); (ii) steel dowels and wires alongside the member (Fig. 11b).

---

[1]For further information on prosthesis, see Chapter "Reinforcement of Historic Timber Roofs" in this RILEM state-of-the-art report.

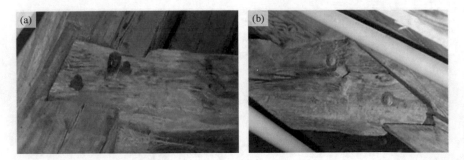

Fig. 10 Old traditional reinforcement of dovetailed half-lap joints using **a** wooden pegs or **b** metal fasteners (bolts, nails or screws) in case of timber splitting [46]

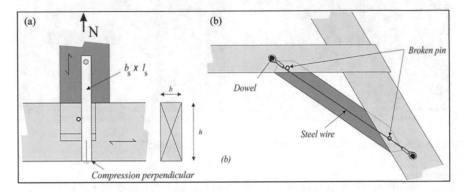

Fig. 11 External reinforcement in tension of mortise-tenon joints featuring broken wooden pins with external metal straps **a** or steel wires **b** fastened with steel dowels to adjoining timber members [46]

The use of screws is another traditional reinforcement technique that has evolved over the ages. In the last two centuries, more accurate thread shaping and a wide range of screw geometry has been made possible, thanks to the emergence of modern technologies and industrial production. Screws and mainly self-tapping screws stand out from other contemporary techniques; they are widely used in timber engineering to reinforce structural elements (e.g. notched beams in bending and shear), timber-to-timber connections in compression-shear-tension, and also dowel-type connections [34, 58, 59]. Self-tapping screws: (i) enhance wood mechanical performance, especially in shear and tension, by ensuring ductile failure mechanisms, higher stiffness and higher load-bearing capacity; (ii) secure structural members and connections by preventing the onset and development of timber splitting, which otherwise would have caused a reduction in their mechanical performance. They provide many more

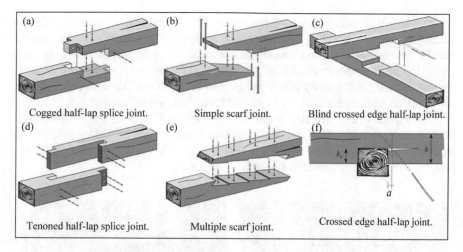

(a) Cogged half-lap splice joint. (b) Simple scarf joint. (c) Blind crossed edge half-lap joint.

(d) Tenoned half-lap splice joint. (e) Multiple scarf joint. (f) Crossed edge half-lap joint.

Fig. 12 Several types of lap and scarf joints reinforced with self-tapping screws [46]

advantages, such as easy handling on-site, low cost, with reduced visual impact, easy visual inspection and high degree of reversibility[2].

As mentioned before, traditional carpentry joints are usually subjected under normal loads to compression, shear, tension and bending (more rarely), and under reverse loads (e.g. winds and seismic events). For every loading configuration, the resulting stress is highly concentrated at the corner of wooden cuts with right or skew angles, which is very typical for carpentry joints with notched zones (i.e. lap, tabled, scarf and step joints). These joints are for this more vulnerable to timber splitting.

Self-tapping screws with partial or full threaded length, can thus be implemented for lap, tabled and scarf joints, under an angle, inclined or perpendicular to the grain (Fig. 12), in order to increase their load-bearing capacity in shear parallel to the grain and tension perpendicular to the grain [46]. As a result, this reinforcement technique reduces the propagation of cracks and counteracts the tendency of timber to split around the notched areas. Furthermore, recent experimentations [55, 56], showed that self-tapping screws could also be used to efficiently reinforce singe step joints to account for compressive crushing at the front-notch surface and shear crack in the tie beam-end.

Lap and scarf joints are usually designed to bear axial compression. Nonetheless, they may also work in tension if they have been reinforced beforehand with metal fasteners (e.g. nails, bolts, screws) perpendicular to their overlap contact surfaces. Indeed, the reinforcement technique consists in making a "dowel-type" connection, for which the steel fasteners work in shear on the overlap area and thus channel axial tensile loads are transferred from both timber adjoining pieces. On the other hand, the "dowel-type" connection is often used for quick and easy repair prosthesis methods, by connecting the new structural members to the existing ones. For this purpose,

---

[2]For further information on self-tapping screws, see Chapter "Self-tapping Screws as Reinforcement for Structural Timber Elements" in this RILEM state-of-the-art report.

half-lap splice joints or simple scarf joints can be designed and reinforced, either by steel fasteners as "dowel-type" connections in shear (Fig. 13a) [57], or by external steel plates fastened on their lateral sides parallel to the overlap surface (Fig. 13b) [46]. The latter could be used if higher tensile stiffness and load-bearing capacity is required for the carpentry joints. As an alternative to the "Trait-de-Jupiter" joint (Fig. 14), glued-in rods can be implemented for a "dowel-type" connection, like metal fasteners placed perpendicular to the overlap surface. However, this reinforcement technique is quite laborious and onerous, since it requires pre-drilled holes inside the connection and significant resin use in situ, the latter depending on the degradation level of the timber members that need to be repaired.

Fig. 13 Lap joints linking the existing timber members with the new ones (prosthesis) by means of bolts (**a**) [57], or by external steel plates fastened with bolts on the overlapping area (**b**) [46]

Fig. 14 Scarf joint (Trait-de-Jupiter) reinforced with steel rods glued with epoxy resin inside both adjoining timber members. *Credits* Pascal Lemlyn (Restoration du Moulin de l'abbaye de la Paix Dieu, Institut du patrimoine Wallon, Belgique)

3.4   Performance Criteria and Applicability Areas

This section focuses on the performance criteria and application areas of reinforcement techniques previously discussed for each typology of traditional carpentry joints. Six criteria for performance and use conditions have been defined in Table 2: (i) implementation; (ii) maintenance; (iii) accessibility; (iv) visual crowding; (v) reversibility; (vi) failure modes. Again, all the traditional and contemporary techniques previously described for the reinforcement of traditional carpentry joints have been sorted into three groups: (i) dowel-type; (ii) inserts; (iii) elements set. The "dowel-type" and "inserts" groups belong to the category of "internal intervention techniques" while "elements set" refers to the "external intervention techniques".

The implementation and maintenance of the "dowel-type" group are quite easy. This is especially true for self-tapping screws, since they can tap their own hole when driven into timber and be removed without difficulty due to their full or partial thread lengths. Apart from skilled carpenters, the maintenance of wooden pegs may nevertheless be challenging for other experts. Although large workspace is optional, the implementation and maintenance of fasteners still require, at least, sufficient accessibility to two parallel faces of the connected timber members for their reinforcement. "Dowel-type" techniques are reversible, almost undetectable, and they provide ductile failures of the connection (e.g. plastic hinges). However, brittle failures such as timber splitting in the vicinity of fasteners may occur when the minimal distance and spacing recommendations are not met.

The implementation of the "inserts" group is quite hard. It requires sufficient workspace and skill to pre-drill or to hollow out connected timber members before reinforcing them with glued-in elements. The maintenance is challenging, but practically infeasible when the control process requires taking out elements that have been glued inside timber. Except for wooden wedges or other unglued pieces, "inserts" techniques are irreversible since they cannot be removed without causing damage to adjoining timber pieces. On the other hand, they are almost undetectable, like most of the reinforcement techniques of the "dowel-type" group. Although ductile failures occur within glued-in elements, timber splitting may also take place at the interface

Table 2 Performance criteria and use conditions for internal and external intervention techniques

Performance criteria and use conditions	Internal interventions		External interventions
	Dowel-type	Inserts	Elements set
Implementation	Easy	Hard	Easy/hard
Maintenance	Easy/hard	Hard	Easy/hard
Accessibility	Low	Low/high	High
Visual crowding	No	No	Yes
Reversibility	Yes	No	Yes/no
Failure modes	Ductile/brittle	Ductile/brittle	Ductile/brittle

between timber and glue, if not spacing requirements are not met, causing significant damage to the members of the connection.

Concerning the external intervention group, "elements set", the implementation and maintenance of fastened elements (e.g. dowelled steel wires, plates or brackets) appear to be easy. Nonetheless, the implementation and maintenance of glued-on elements can be a bit more challenging for the same reasons as previously stated for the "inserts" group. In contrast to both internal intervention groups, reinforcement techniques belonging to "elements set" require a very large work area, mostly depending on the intervention extent, and skilled expert able to manipulate different materials in a systemic whole. The external interventions, may result in strong visual crowding on the aesthetical aspect of timber structures. In addition, their reversibility can be discussed, depending on the techniques and materials used to reinforce the timber connections. Despite this, "elements set" techniques often provide higher tensile and shear strength to timber connections, as well as failure modes more ductile than those obtained from the "dowel-type" or "inserts" groups. Nevertheless, timber splitting may also occur and should thus be prevented when designing reinforced connections, such as for both internal intervention groups. Lastly, transverse dimensional changes to the grain in timber members due to moisture content fluctuations should also be taken into account, since steel elements have a completely different behaviour, preventing these movements and causing splitting in timber. Moisture condensation on steel surfaces is another parameter that needs to be taken into account when deciding an intervention. Based on the performance criteria and use conditions, some information about the applicability areas of internal and external reinforcement intervention techniques is presented in Table 3, for the six typologies of traditional carpentry joints described in Sect. 2.1. Thus, the "dowel-type" reinforcement techniques can easily be applied to carpentry joints, except for some heading joints with complex geometrical configuration at their joined ends. The respective applicability of the intervention techniques under the "inserts groups" depends on the initial geometry of the traditional carpentry joint. The higher the amount of notched areas within the connection, the harder the applicability of such intervention techniques. This is especially true for glued-in elements, for which pre-drilling or

Table 3 Applicability of internal and external intervention techniques to traditional carpentry joints

Typologies of traditional carpentry joints	Internal interventions		External interventions
	Dowel-type	Inserts	Elements set
Lap joints	A	A/NA	A
Tabled joints	A	A/NA	A
Scarf joints	A	A/NA	A
Heading joints	A/NA	A/NA	A
Mortise-tenon joints	A	A/NA	A
Step joints	A	A	A

Legend A—applicable; NA—not applicable

hollowing out of wood is required in adjoining members beforehand, more often at their contact surfaces, causing significant damage to the carpentry connections and reducing their mechanical performance. On the other hand, external intervention techniques can be carried out either at a nodal scale in every traditional carpentry joint or at a structural scale in timber members. With a wide applicability area for every traditional carpentry joint, the "elements set" group presents advantages, such as great reversibility, better enhancement of the mechanical performance and lower risk of damage to existing timber structures during the implementation stages of the reinforcement techniques.

4 Conclusions

This chapter has aimed to present state-of-the-art reinforcement techniques of traditional timber carpentry joints by collating information from available standards, numerical investigations, in situ case studies and lab experiments. For better understanding how these connections work within timber structures and their failure mechanisms, such as compressive crushing, shear and tensile cracks, six geometrical typologies of traditional carpentry joints have been discussed. Based on this knowledge, reinforcement strategies were presented, with greater focus on their objectives, methodologies, techniques, performance criteria and applicability areas.

Many traditional and contemporary techniques exist for the reinforcement of traditional carpentry joints. A suitable reinforcement technique, based on strongly defined methodologies, should be wisely selected on a case-by-case study by considering the following factors: (i) geometry and typology of the traditional carpentry joint that needs to be reinforced; (ii) type and extent of defects and damage of the adjoining timber members; (iii) conservation level of the investigated structure; (iv) accessibility and on-site conditions; (v) feasibility and execution process; (vi) visual crowding and reversibility.

Other challenges that can be emphasized when assessing and reinforcing traditional carpentry joints, are:

- Both stiffness and strength of joints highly depend on the reinforcement methodology and technique applied, but also on the type of loading (e.g. monotonic, cyclic, in or out of plane, etc.). For instance, the rotational stiffness of reinforced traditional carpentry joints is usually different under positive and negative bending [50–53]. Moreover, for most connections, there is an interaction between the different pathways in which the forces are transferred in terms of stiffness and strength. Changes in joint stiffness may have severe consequences at the overall behaviour and load distribution among the members of the entire timber structure, altering the paths of loads, transferring loads and problems to other areas. This interaction should be considered to better define the mechanical performance of traditional carpentry joints and their respective reinforcement strategies.

- An important parameter for the load bearing capacity of traditional carpentry joints is the tight fitting and contact between the assembled elements, which especially in traditional timber structures is not always met for many reasons. Though several experimental and numerical studies have already been carried out in the last decades on carpentry joints, it is not easy to calculate the value of the contact pressure in the following situations: (i) unknown contact surfaces, namely undetermined amount of effective contact surfaces and respective load-transfer contributions; (ii) non-uniform stress distributions, triggered by the geometry of the joint, the presence of local wood defects (e.g. drying cracks, knots, etc.) and the differences of craftsmanship that can be found even in the same roof.
- Although some rules for the assessment and reinforcement of the material strength of timber members (e.g. beams, columns, etc.) are provided by most current Standards and National Annexes [25, 30, 33], their extension to traditional carpentry joints is an objective that hopefully would be attained in future, in order to give to the involved professionals (architects, structural engineers, carpenters), the knowledge and tools to repair or reinforce them, with techniques and methods that respect and preserve the function and authenticity of the timber structures, while ensuring a sufficient level of safety.

References

1. Goss WFM (1890) Bench work in wood. A course of study and practice designed for the use of schools and colleges. Ginn & Company, Boston, USA. http://www.woodworkslibrary.com/repository/bench_work_in_wood.pdf
2. Oslet G (1890) Traité de charpente en bois. Encyclopédie théorique & pratique des connaissances civiles et militaires. In: Fils CH (ed) Partie Civile, Cours de construction, Quatrième partie. Digital reproduction, Paris, France
3. Grezel J (1950) Les assemblages. Annales de l'Institut Techniques du Bâtiment et des Travaux Publics, Manuel de la charpente en bois, n°9. Institut Technique du Bâtiment et des Travaux Publics, Novembre 1950, Paris, France
4. Rogowski G (2002) The complete illustrated guide to joinery. The Taunton Press, Inc.
5. Sobon JA (2004) Historic American timber joinery. In: Rower K (ed) A graphic guide. National Center for Preservation Technology and Training, USA, 57 Pages
6. Hirst E, Brett A, Thomson A, Walker P, Harris R (2008) The structural performance of traditional oak tension & scarf joints. In: WCTE—proceedings of the 10th world conference on timber engineering. Miyazaki, Japan. June 2–5
7. Arciszewska-Kedzior A, Kunecky J, Hasníková H, Václav Sebera V (2015) Lapped scarf joint with inclined faces and wooden dowels: experimental and numerical analysis. Eng Struct 94:1–8. https://doi.org/10.1016/j.engstruct.2015.03.036
8. Perria E (2016) Characterization of halved undersquinted scarf joint and stop-splayed undersquinted and tabled scarf joint with key (Jupiter joint). PhD thesis. University of Florence, Italy. University of Braunschweig, Germany, 305 Pages
9. Fajman P, Máca J (2018) Stiffness of scarf joints with dowels. Comput Struct 207:194–199. https://doi.org/10.1016/j.compstruc.2017.03.005
10. Sangree RH, Schafer BW (2009) Experimental and numerical analysis of a halved and tabled traditional timber scarf joint. Constr Build Mater 23(2):615–624. https://doi.org/10.1016/j.conbuildmat.2008.01.015

11. Aira JR, Arriaga F, Íñiguez-González G, Guaita M (2015) Failure modes in halved and tabled tenoned timber scarf joint by tension test. Constr Build Mater 96:360–367. https://doi.org/10.1016/j.conbuildmat.2015.08.107

12. Aira JR, Íñiguez-González G, Guaita M, Arriaga F (2016) Load carrying capacity of halved and tabled tenoned timber scarf joint. Mater Struct 49:5343–5355. https://doi.org/10.1617/s11527-016-0864-y

13. Sainju PR, Shrestha M, Prajapati R, Prajapati RS, Shrestha R (2019) Performance of timber joint system used in traditional monuments of Bhaktapur municipality. 2nd international conference on earthquake engineering and post disaster reconstruction planning. Bhaktapur, Nepal. April 25–27, 2019

14. Crayssac E, Song X, Wu Y, Li K (2018) Lateral performance of mortise-tenon jointed traditional timber frames with wood panel infill. Eng Struct 161:223–230. https://doi.org/10.1016/j.engstruct.2018.02.022

15. Xie Q, Zhang L, Wang L, Zhou W, Zhou T (2019) Lateral performance of traditional Chinese timber frames: experiments and analytical model. Eng Struct 186:446–455. https://doi.org/10.1016/j.engstruct.2019.02.038

16. Sandberg LB, Bulleit WM, Reid EH (2000) Strength and stiffness of oak pegs in traditional timber-frame joints. J Struct Eng 126(6):717–723

17. Shanks J, Walker P (2005) Experimental performance of mortice and tenon connections in Green Oak. Struc Eng 83(17)

18. Likos E, Haviarova E, Eckelman CA, Erdil YZ, Ozcifci A (2012) Effect of tenon geometry, grain orientation, and shoulder on bending moment capacity and moment rotation characteristics of mortise and tenon joints. Wood Fiber Sci 44(4):462–469

19. Hajdarević S, Martinović S (2014) Effect of tenon length on flexibility of mortise and tenon joint. Proc Eng 69:678–685. https://doi.org/10.1016/j.proeng.2014.03.042

20. Parisi MA, Cordié C (2010) Mechanical behaviour of double-step timber joints. Const Build Mater 24(8):1364–1371

21. Feio AO, Lourenço PB, Machado JS (2014) Testing and modelling of a traditional timber mortise and tenon joint. Mater Struct 47(1–2):213–225

22. Verbist M, Branco JM, Poletti E, Descamps T, Lourenço PB (2017) Single step joint: overview of European standardized approaches and experimentations. Mater Struct 50:161–2017. https://doi.org/10.1617/s11527-017-1028-4

23. Villar-García JR, Crespo J, Moya M, Guaita M (2018) Experimental and numerical studies of the stress state at the reverse step joint in heavy timber trusses. Mater Struct 51:17. https://doi.org/10.1617/s11527-018-1144-9

24. Villar-García JR, Vidal-López P, Crespo J, Guaita M (2019) Analysis of the stress state at the double-step joint in heavy timber structures. Materiales de Construcción 69(335). https://doi.org/10.3989/mc.2019.00319

25. EN 1995-1-1 (2004) Eurocode 5—design of timber structures—part 1.1: general—common rules and rules for buildings. CEN, European Standardisation Institute, Brussels, Belgium

26. Siem J, Jorissen A (2015) Can traditional carpentry joints be assessed and designed using modern standards? Structural health assessment of timber structures. In: Jerzy J, Tomasz N (eds) Shatis'15: 3rd international conference on structural health assessment of timber structure, vol 1. Wroclaw, Poland, pp 108–119

27. de Rijk R, Jorissen A (2016) In: Jorissen AJM, Leijten AJM (eds) Spanningen in een Tand-verbinding. Onderzoeksmiddag construeren met hout: 22 januari 2016. Technische Universiteit Eindhoven, Nederland, pp 42–69. ISBN: 978-90-386-3993-2

28. Aira JR, Descamps T, Van Parys L, Léoskool L (2015) Study of stress distribution and stress concentration factor in notched wood pieces with cohesive surfaces. Eur J Wood Wood Prod 73(3):325–334

29. Branco JM, Verbist M, Descamps T (2018) Design of three step joint typologies: review of European standardized approaches. Eng Struct 174:573–585. https://doi.org/10.1016/j.engstruct.2018.06.073

30. EN 1995-1-1/A1 (2008) Eurocode 5—amendment 1—design of timber structures—part 1.1: general—common rules and rules for buildings. CEN, European Standardisation Institute. November 26, 2008, Brussels, Belgium
31. Sandberg D, Söderström O (2006) Crack formation due to weathering of radial and tangential sections of pine and spruce. Wood Mater Sci Eng 1(1):12–20. https://doi.org/10.1080/174802 70600644407
32. Mergny E, Mateo R, Esteban M, Descamps T, Latteur P (2016) Influence of cracks on the stiffness of timber structural elements. In: WCTE 2016—proceedings of the world conference on timber engineering, Vienna, Austria
33. EN 1998-1 (2004) Eurocode 8: design of structures for earthquake resistance—part 1: general rules. CEN, European Standardisation Institute. Brussels, Belgium
34. Lathuillière D, Bléron L, Descamps T, Bocquet J-F (2015) Reinforcement of dowel type connections. Constr Build Mater 97:48–54. https://doi.org/10.1016/j.conbuildmat.2015.05.088
35. Uibel T, Blass HJ (2010) A new method to determine suitable spacings and distances for self-tapping screws. International council for research and innovation in building and construction—working commission W18, timber structures. New Zealand
36. ETA-11/0030, European Technical Approval (2012) Self-tapping screws for use in timber structures. EOTA—European Organization for Technical Approval. Rotho Blaas Self-tapping Screws
37. Freas A (1982) Evaluation, maintenance and upgrading of wood structures—a guide and commentary. American Society of Civil Engineers (ASCE), New York, USA
38. UNI 11119 (2004) Cultural heritage—wooden artefacts—load bearing structures—on site inspections for the diagnosis of timber members. Ente Nazionale Italiano di Unificazione, Milano
39. Cruz H, Yeomans D, Tsakanika E, Macchioni N, Jorissen A, Touza M, Mannucci M, Lourenço PB (2015) Guidelines for on-site assessment of historic timber structures. Int J Archit Heritage 9(3):277–289. https://doi.org/10.1080/15583058.2013.774070
40. Riggio M, Tannert T (2016) Structural assessment: diagnosis, before intervention! In: Eberhard-steiner J, Winter W, Fadai A, Pöll M (eds) World conference on timber engineering—WCTE 2016. Vienna, Austria. August 22–25, 2016. ISBN: 978-3-903039-00-1
41. ICOMOS Charter (1999) Principles for the preservation of historic timber structures. International council on monuments and sites—ICOMOS. 12th general assembly. Mexico City, Mexico, October 1999
42. Russell R (2013) Structural timber repairs. The building conservation directory. http://www.buildingconservation.com/articles/structural-timber-repairs.htm
43. Croatto G, Turrini U (2014) In: Sousa HS, Branco JM, Lourenço PB (eds) Restoration of historical timber structures—criteria, innovative solutions and case studies. Intervir em construções existentes de madeira—Livro de Atas. Guimarães, Portugal, June 5, pp 119–136
44. Tampone G, Ruggieri N (2016) State-of-the-art technology on conservation of ancient roofs with timber structure. J Cult Heritage 22:1019–1027. https://doi.org/10.1016/j.culher.2016.05.011
45. Branco JM (2014) Reforço de elementos existentes de madeira. In: Sousa HS, Branco JM, Lourenço PB (eds) Intervir em construções existentes de madeira—Livro de Atas. Guimarães, Portugal, June 5, pp 57–70
46. Branco JM, Descamps T (2015) Analysis and strengthening of carpentry joints. Constr Build Mater 97:34–47. Special Issue: Reinforcement of Timber Structures. https://doi.org/10.1016/j.conbuildmat.2015.05.089
47. Munafò P, Stazi F, Tassi C, Davi F (2015) Experimentation on historic timber trusses to identify repair techniques compliant with the original structural-constructive conception. Constr Build Mater 87:54–66
48. Wheeler AS, Hutchinson AR (1998) Resin repairs to timber structures. Int J Adhes Adhes 8:1–13
49. Descamps T, Avez C, Carpentier O, Antczak E, Jeong GY (2016) Historic timber roofs modelling: prosthesis and resin repairs. J Heritage Conser 47:52–60. https://doi.org/10.17425/WK47ROOFS

50. Branco JM, Piazza M, Cruz PJS (2011) Experimental evaluation of different strengthening techniques of traditional timber connections. Eng Struct 33(8):2259–2270. https://doi.org/10.1016/j.engstruct.2011.04.002

51. Palma P, Garcia H, Ferreira J, Appleton J, Cruz H (2012) Behaviour and repair of carpentry connections—rotational behaviour of the rafter and tie beam connection in timber roof structures. J Cult Heritage 13(3):S64–S73. https://doi.org/10.1016/j.culher.2012.03.002

52. Parisi MA, Piazza M (2015) Seismic strengthening and seismic improvement of timber structures. Constr Build Mater 97:55–66. https://doi.org/10.1016/j.conbuildmat.2015.05.093

53. Drdácký M, Urushadze S (2019) Retrofitting of imperfect halved dovetail carpentry joints for increased seismic resistance. Buildings 9(2):12. https://doi.org/10.3390/buildings9020048

54. Zhao X-b, Zhang F-l, Xue J-y, Ma L-l (2019) Shaking table tests on seismic behaviour of ancient timber structure reinforced with CFRP sheet. Eng Struct 197:16. https://doi.org/10.1016/j.engstruct.2019.109405

55. Sobra K, de Rijk R, Aktas YD, Avez C, Burawska I, Branco JM (2016) Experimental and analytical assessment of the capacity of traditional single notch joints and impact of retrofitting by self-tapping screws. In: Cruz H et al (eds) Historical earthquake-resistant timber framing in the mediterranean area, pp 359–369. https://doi.org/10.1007/978-3-319-39492-3_30

56. Verbist M, Branco JM, Poletti E, Descamps T, Lourenço PB (2018) Experimentations on the retrofitting of damaged single step joints with self-tapping screws. Mater Struct 51:106. https://doi.org/10.1617/s11527-018-1234-8

57. Pinto L (2008) Inventory of repair and strengthening methods—timber. Advanced masters in structural analysis of monuments and historical constructions. Master thesis, University of Catalonia, Spain

58. Dietsch P, Brandner R (2015) Self-tapping screws and threaded rods as reinforcement for structural timber elements—a state-of-the-art report. Constr Build Mater 97:78–89. https://doi.org/10.1016/j.conbuildmat.2015.04.028

59. Franke S, Franke B, Harte AM (2015) Failure modes and reinforcement techniques for timber beams—state of the art. Constr Build Mater 97:2–13. https://doi.org/10.1016/j.conbuildmat.2015.06.021

Printed in the United States
by Baker & Taylor Publisher Services